A level in a week

Biology

D0756494

Jim Sharpe and Kevin Byrne, Abbey Tutorial College
Series editor: Kevin Byrne

Where to find the information you need

Letts Educational
Aldine Place
London W12 8AW
Tel: 0181 740 2266
Fax: 0181 743 8541
e-mail: mail@lettsed.co.uk
website: http://www.lettsed.co.uk

First published 1999
Reprinted 1999

British Library Cataloguing in Publication Data
A CIP record for this book is available from the British Library.

ISBN 1 85758 925 4

Editorial, design and production by Hart McLeod, Cambridge

Printed in Great Britain

Letts Educational is the trading name of BPP (Letts Educational) Ltd

Cells and membranes

20 minutes

1 Organisms can be classified as viruses, _____ and eukaryotes. The smallest of these are the _____ which, because they are totally dependent upon a host cell, are known as _____ _____.

2 The cell surface membrane has a basic structure known as the _____ _____ model. The membrane is _____ wide and is composed of a _____ bilayer with proteins.

3 Substances may enter or leave cells by diffusion, osmosis, facilitated diffusion and _____ _____. This process requires energy in the form of _____.

4 The nucleus encloses and protects the _____, which controls the activities of the cell. The nuclear envelope contains pores which allow mRNA to _____ the nucleus during protein synthesis.

5 The inner membrane of the mitochondria is folded into _____ and is the site of the _____ _____ _____. The matrix is the site of the _____ cycle.

6 The rough endoplasmic reticulum (ER) is covered in _____ and synthesized proteins can be moved through the cell in its cavities. Smooth ER is a site of _____ metabolism.

7 The rough endoplasmic reticulum is connected to flat cisternae known as the _____ _____. Vesicles pinched off from this can be used to transport substances such as enzymes for _____.

8 The largest organelles in plant cells are _____. Stroma are the site of the _____ _____ _____ whilst the light-dependent stage occurs in the _____ .

Answers

1 prokaryotes, viruses, obligate parasites 2 fluid mosaic, 7.5 nm, phospholipid 3 active transport, ATP 4 DNA, leave 5 cristae, electron transport chain, Krebs 6 ribosomes, lipid 7 golgi body, secretion 8 chloroplasts, light-independent stage, thylakoids

If you got them all right, skip to page 6

Cells and membranes

Improve your knowledge

1 Organisms can be classified as **viruses**, **prokaryotes** and **eukaryotes**.

	Viruses	Prokaryotes	Eukaryotes
Example	Flu virus, HIV	Bacteria	Plants, animals, fungi, protoctists
Size	10–30 nm (nanometres)	0.1–10 μm (micrometres)	10–100 μm (micrometres)
Nuclear material	Either RNA or DNA plus proteins	Circular DNA, no histone proteins. Not organised within a membrane-bound nucleus	Linear DNA attached to histone proteins. Organised within a membrane-bound nucleus
Internal organisation	No membranes or organelles	No membranes and organelles	Complex membranes compartmentalize the cell

2 The **membranes** which surround eukaryotic cells and the membranes which form their organelles all have the same basic structure, known as the **fluid mosaic model**. It is 7.5 nm thick and consists of a **lipid bilayer** with some proteins imbedded in the surface and some running right through it.

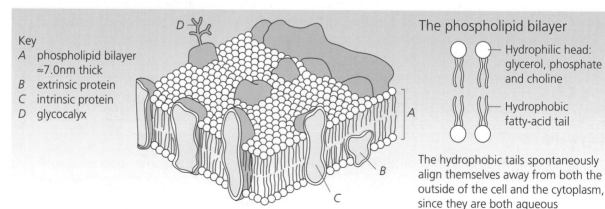

Key
A phospholipid bilayer ≈7.0nm thick
B extrinsic protein
C intrinsic protein
D glycocalyx

The phospholipid bilayer

Hydrophilic head: glycerol, phosphate and choline

Hydrophobic fatty-acid tail

The hydrophobic tails spontaneously align themselves away from both the outside of the cell and the cytoplasm, since they are both aqueous

The **interaction** between the hydrophobic and hydrophilic ends of the phospholipids gives the membranes **stability**. **Lipids** also give the membranes **selective permeability**. Lipid-soluble (hydrophobic) molecules diffuse through the membrane easily. Hydrophilic substances cross the membrane via water-filled pores or through channels in the intrinsic proteins. The lipids/proteins can move laterally or change places and this gives the membrane fluidity. This is essential for processes such as endocytosis (movement of material into a cell).

3 Substances may cross the plasma membrane to enter or leave cells by four mechanisms.

Mechanism	Description	ATP required	Example
Diffusion	Movement of molecules from a region of high concentration to low concentration	No	Oxygen, carbon dioxide
Osmosis	Diffusion of H_2O	No	Water
Facilitated diffusion	Movement of substances by attachment to transport proteins	No	Glucose
Active transport	Movement of substance against their concentration gradient	Yes	Glucose

4 Within eukaryotic cells the cytoplasm is divided up by membranes into compartments or organelles. Each organelle is specialised for a particular function.

Organelle	Structure and function
Nucleus Nuclear pore, Nuclear envelope, Nucleoplasm containing chromatin, Nucleolus	A double nuclear envelope encloses and protects **DNA** (normally visible as chromatin granules).
	Nuclear pores allow entry of substances such as nucleotides for DNA replication and exit of molecules such as **mRNA** during protein **synthesis**.
	The outer membrane of the nuclear envelope is continuous with the rough endoplasmic reticulum membranes. This makes the perinuclear space (space within the nuclear envelope) continuous with the lumen of the endoplasmic reticulum, thus allowing easy transport of substances.

(table continues on page 4)

	Organelle	Structure and function
5	**Mitochondria** Crista, Matrix, Ribosomes, Fluid-filled space, Loop of DNA, Inner membrane, Outer membrane	A double membrane isolates reactions of the Krebs cycle and electron transport chain from the general cytoplasm. Such compartmentalization allows high concentrations of enzymes and substrates to be maintained which increases the rate of respiratory reactions.
		The inner membrane is folded to form **cristae** which greatly increase the surface area for the electron transport chain reactions.
		The matrix contains • 70S ribosomes for protein manufacture, DNA for protein manufacture, and • Enzymes, e.g. decarboxylase used in Krebs cycle.
6	**Endoplasmic reticulum**	Endoplasmic reticulum is a system of hollow tubes and sacs which allow transport of substances within the cell.
		Rough endoplasmic reticulum (RER) is covered with ribosomes and consists of an interconnected system of flattened sacs. The ribosomes synthesise proteins which can then be transported through the cell in the cavities of the endoplasmic reticulum. The percentage of RER is high in cells which produce proteins for export, e.g. digestive enzymes.
		Smooth endoplasmic reticulum (SER) – which lacks ribosomes – is a system of interconnected tubules. This is the site of carbohydrate and lipid metabolism.
7	**Golgi body**	The Golgi body consists of flattened **cisternae** (membrane bound cavities) which allows internal transport. Vesicles contain materials to be secreted.
		The Golgi body is connected to the RER. Proteins from the RER are modified before secretion. For example, carbohydrates may be added to proteins to form glycoproteins such as mucus.

(table continues on page 5)

Organelle	Structure and function
Ribosomes 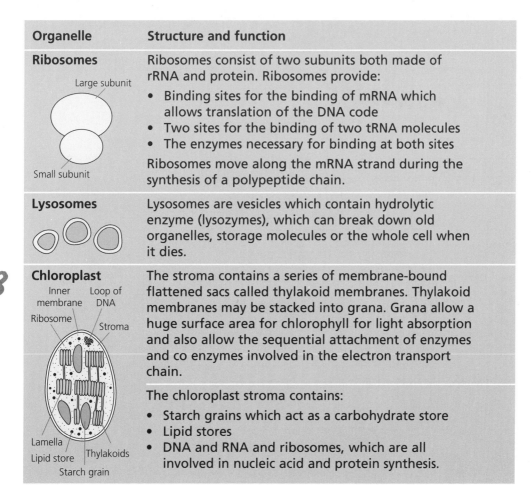	Ribosomes consist of two subunits both made of rRNA and protein. Ribosomes provide: • Binding sites for the binding of mRNA which allows translation of the DNA code • Two sites for the binding of two tRNA molecules • The enzymes necessary for binding at both sites Ribosomes move along the mRNA strand during the synthesis of a polypeptide chain.
Lysosomes	Lysosomes are vesicles which contain hydrolytic enzyme (lysozymes), which can break down old organelles, storage molecules or the whole cell when it dies.
Chloroplast	The stroma contains a series of membrane-bound flattened sacs called thylakoid membranes. Thylakoid membranes may be stacked into grana. Grana allow a huge surface area for chlorophyll for light absorption and also allow the sequential attachment of enzymes and co enzymes involved in the electron transport chain.
	The chloroplast stroma contains: • Starch grains which act as a carbohydrate store • Lipid stores • DNA and RNA and ribosomes, which are all involved in nucleic acid and protein synthesis.

8

Cells and membranes

Use your knowledge

1 Complete the table below which compares cellular transport mechanisms.

	Simple diffusion	Facilitated diffusion	Active transport
Is ATP required?	N		Y
Rate of movement		Fast	
Direction of transport in relation to concentration gradient	With		Against

2 The diagram shows part of a cell surface membrane.

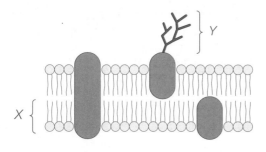

(a) Identify structure:
    (i) X
    (ii) Y.
(b) Explain why the cell surface membrane is said to be *fluid*.
(c) Suggest why some liver cells have very high percentages of
    (i) SER
    (ii) Mitochondria.

Answers on page 91

Enzymes and respiration

20 minutes

1 Enzymes are biological catalysts, reducing the _____ energy of reactions. They are composed of _____ proteins with a specific tertiary structure.

2 When a substrate enters an enzyme's active site, the enzyme and substrate form a _____ relationship. In the _____ _____ _____ hypothesis substrate shape is complementary to the shape of the active site.

3 Increasing temperature increases the _____ energy of enzyme and substrate molecules, with high temperature causing the enzyme to become irreversibly _____. Enzymes operate in a _____ pH range, with changing pH around the enzyme disrupting the _____ _____ _____ _____ _____.

4 Increasing enzyme concentration causes a _____ increase in the rate of an enzyme-catalysed reaction, whereas increasing substrate concentration also increases rate, but only until a _____ _____ is reached.

5 Competitive inhibitors bind to the enzyme's _____ site, whereas _____ inhibitors bind elsewhere on the enzyme.

6 During glycolysis, glucose is _____ using two ATP molecules, to increase its reactivity. One molecule of glucose in glycolysis generates two molecules of _____, two molecules of $NADH_2$ and two _____ overall.

7 In the presence of O_2, pyruvate combines with coenzyme A to form _____ _____. This enters the _____ cycle and combines with oxaloacetic acid to form _____.

8 The electron transport chain is a series of carrier molecules through which pairs of _____ are transferred, releasing energy to convert _____ to ATP. This process is called _____ _____. The total yield in aerobic respiration is _____ ATP molecules/glucose molecule.

Answers

1 activation, globular 2 steric, lock and key 3 kinetic, denatured, narrow, shape of the active site 4 linear, maximum rate 5 active, non-competitive 6 phosphorylated, pyruvate, ATP 7 acetyl coenzyme A, Krebs, citrate 8 electrons, ADP, oxidative phosphorylation, 38

✓ *If you got them all right, skip to page 14*

Enzymes and respiration

Improve your knowledge

20 minutes

1   Enzymes control metabolic reactions in the body and the rate at which
    they occur. Metabolic reactions can be **catabolic** (breakdown reactions,
    which release energy overall), or **anabolic** (building up or synthesising
    reactions, requiring energy overall).

Enzymes have the following characteristics:

- **Biological catalysts**, made up of globular proteins with a specific
  tertiary structure.
- They reduce the **activation energy** of reactions, which is the energy
  required for a specific reaction to occur.
- They **speed up** and **slow down** metabolic reactions but are never used
  up in the reaction.

Tertiary structure is the 3D folding of protein chains

2   The **active site** is a depression in an enzyme's tertiary structure and is the
    site of the enzyme-catalysed reaction. The shape of the active site allows
    only specific substrate molecules to enter, which results in **enzyme
    specificity**, with enzymes unique to a specific reaction. The enzyme and
    substrate are said to form a **steric relationship**.

There are two theories about the action of enzymes:

- **Lock and key** – Substrate shape is directly complementary to the
  shape of the active site, with the enzyme acting as the key and the
  substrate molecules the lock.
- **Induced fit** – Not all enzymes have a permanent active site. In some,
  one develops as substrate molecules come close, with a small change
  occurring in the enzyme structure to form a specific active site.

3   The rate of enzyme-catalysed reactions will increase with increasing temperature, up to an optimum rate. Above this optimum, further temperature increase decreases the rate of reaction.

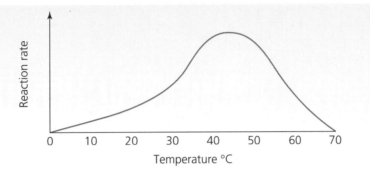

Increasing temperature increases the enzyme and substrate molecule's **kinetic energy**. The faster-moving molecules collide more often, forming more enzyme–substrate complexes. Temperatures above optimum cause increasing enzyme molecule vibration, breaking down internal bonds and destroying the active site. The enzyme becomes irreversibly **denatured**.

Formation of more enzyme–substrate complexes results in an increased rate of reaction

Enzymes are very **sensitive to pH**. Individual enzymes operate in a narrow pH range and either side of this optimum irreversible denaturation will occur. Optimum pH for most enzymes is 7 (neutral), although some digestive enzymes, e.g. pepsin, have their optimum rate in acidic conditions.

Changing pH causes **changes** in the **charge** of an enzyme's **amino acid components**. This alters attraction and repulsion forces within the enzyme disrupting the shape of the active site.

Remember, the stomach has a pH of 1.5. Some stomach enzymes are adapted to this

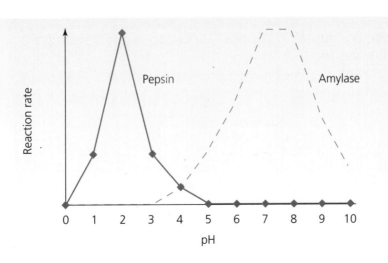

4 The rate of an enzyme-catalysed reaction is also affected by concentration of both enzyme and substrate.

Enzyme concentration	Increasing enzyme concentration causes a **linear increase** in rate (provided the substrate concentration is not limiting)	Increasing enzyme concentration increases the number of active sites available to catalyse the reaction

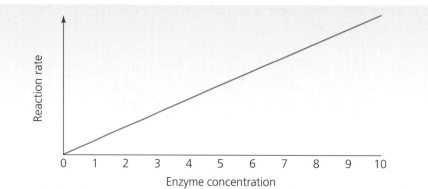

Substrate concentration	Increasing substrate concentration increases rate until a **maximum rate** is reached (plateau)	Maximum rate is due to all available active sites becoming occupied by substrate molecules. Increasing substrate molecules will therefore not increase the rate

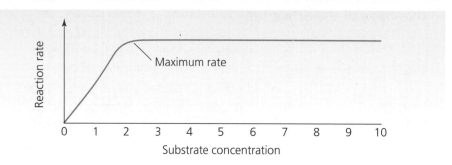

5 **Inhibitors** are substances that prevent normal enzyme activity.

- **Competitive inhibitors** compete with the substrate for the active site. The level of inhibition is dependent upon the relative concentrations of substrate and inhibitor, with increasing inhibitor concentration increasing the level of inhibition. This type of inhibition is **reversible**.
- **Non-competitive inhibitors** bind to the enzyme, but not at the active site. The level of inhibition depends only on the concentration of the inhibitor, not the substrate, because the inhibitor has a higher affinity for the enzyme than the substrate. This type of inhibition is **irreversible**.

Heavy metals such as arsenic and mercury are non-competitive inhibitors, binding to sulphur groups present

6 Cellular respiration is an example of a sequence of enzyme controlled reactions, involving the **breakdown of complex molecules**, e.g. glucose and fatty acids, to simpler molecules to **release energy** in the form of **ATP** (adenosine triphosphate).

ATP is the **energy-storage molecule** made during respiration. The stored energy is released when ATP is split into ADP (adenosine diphosphate) by losing a phosphate group. Cells use this energy for metabolic activities, e.g. synthesis of molecules and active transport.

Glycolysis is the first stage of respiration and involves the breakdown of glucose in the cytoplasm to produce pyruvate.

Glycolysis uses two ATP molecules to produce four ATP molecules

Glucose
↓
Fructose–1–6–bisphosphate
↓
Glyceraldehyde–3–phosphate
↓
Pyruvate

- Glucose is **phosphorylated** using phosphate ions from two ATP molecules. This increases reactivity of the glucose forming fructose–1–6–bisphosphate.
- Unstable fructose–1–6–bisphosphate breaks down to two molecules of glyceraldehyde–3–phosphate.
- Each glyceraldehyde–3–phosphate donates a pair of H^+ ions to NAD (NAD is **reduced**) producing $2NADH_2$ molecules, four ATP molecules and pyruvate.

From one molecule of glucose, glycolysis generates two molecules of pyruvate, two $NADH_2$ (reduced NAD) and two ATP overall.

7 During **aerobic respiration** (if oxygen is present) pyruvate enters the mitochondria matrix and is converted to acetate, with one pair of H^+ ions reducing NAD to $NADH_2$ and CO_2 released. Acetate combines with coenzyme A forming the two-carbon molecule **acetyl coenzyme A**. This is known as the **link reaction**.

Acetyl coenzyme A enters the **Krebs cycle** and combines with **oxaloacetic acid** (four-carbon) to form **citric acid** (six-carbon). Citric acid is broken down to oxaloacetic acid, via a series of intermediate steps, releasing four pairs of H^+ ions (three reducing NAD and one reducing FAD), one molecule of ATP, and two molecules of CO_2.

8 The **electron transport chain (ETC)** is a series of co enzymes or carrier molecules of progressively lower energy levels through which the pairs of electrons on reduced NAD and FAD are transported. This transport releases energy allowing ADP to be converted to ATP. Each $NADH_2$ generates three ATP and each $FADH_2$ generates two ATP.

ETC accepts electrons allowing NAD/FAD to be recycled

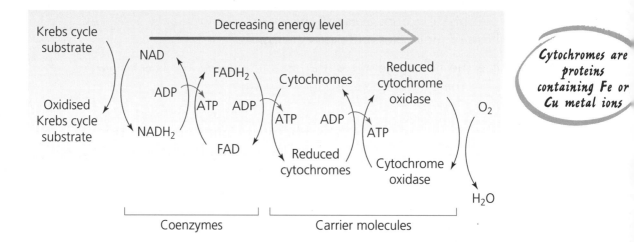

Cytochromes are proteins containing Fe or Cu metal ions

This process is called **oxidative phosphorylation** because pyruvate is oxidised, and ADP is phosphorylated to ATP.

Total yield in aerobic respiration is **38 ATP molecules/glucose molecule**:

* Glycolysis generates two ATP directly and six ATP from two $NADH_2$.
* Krebs cycle and ETC generates 30 ATP.

Oxygen is vital in the ETC as the final acceptor. **Anaerobic respiration** occurs when O_2 is absent, because NAD (and FAD) are not recycled at the ETC, therefore Krebs cycle stops.

During anaerobic respiration in **plant** and **yeast** cells, pyruvate produced in glycolysis is converted to ethanal (with the loss of CO_2). Ethanal is then reduced to **ethanol** using a pair of H+ ions from $NADH_2$ (produced in glycolysis). This is also known as **fermentation**.

During anaerobic respiration in **animal** cells, pyruvate is reduced to **lactic acid** (using a pair of H+ ions from glycolysis).

Since anaerobic respiration is only the partial breakdown of glucose it produces only **two molecules of ATP** from the glycolysis of one molecule of glucose, compared to the 38 ATP from aerobic respiration.

Now learn how to use your knowledge

Enzymes and respiration

Use your knowledge

20 minutes

Hints

1 (a) Explain how the shape of an enzyme molecule enables specificity.
   (b) Name three factors which affect enzyme activity.
   (c) Explain the following terms in relation to enzyme activity:
        (i)   Active site.
        (ii) Competitive inhibition.

How do enzyme reactions occur?

2 An experiment was conducted to investigate the activity of the enzyme catalase, in potatoes. Ground-up potato tissue (with buffer solution) was mixed with a substrate in a conical flask and the volume of oxygen bubbles produced was measured.

   (a) Explain why a buffer solution is mixed with the catalase.
   (b) State two reasons for grinding up the potato tissue.
   (c) Suggest a name for the substrate on which catalase acts.
   (d) State and explain two other pieces of apparatus that should be used when conducting this experiment.

What factors will affect enzyme activity?

3 Cellular respiration can be divided into a series of steps: glycolysis, Krebs cycle and oxidative phosphorylation.

   (a) Name a respiratory substrate that would be used in a liver cell.
   (b) State in which part of the liver cell glycolysis occurs.
   (c) Explain how oxidative phosphorylation produces ATP.

Answers on page 91

Organic molecules

Test your knowledge

1 Water molecules have two charged ends and can form weak _____ bonds. It is said to be a _____ molecule.

2 The basic units of carbohydrates are _____, characterised by the number of carbon atoms, e.g. _____ sugars have five carbons.

3 Two monosaccharides can link together to form a _____. The bond between them is a _____ linkage formed in a _____ reaction.

4 Any disaccharide containing a carbonyl group available to donate electrons can act as a _____ sugar, and will turn _____ solution to an orange-red precipitate.

5 Starch is a polysaccharide composed of _____ and amylopectin. Its main function is as a _____ carbohydrate. Starch and glycogen are suitable for this function because they are _____ and compact.

6 _____ fatty acids are synthesised by the body, whereas _____ fatty acids must be obtained in the diet. Lipids function as a metabolic fuel store as they contain double the amount of energy of _____.

7 Fatty acids with double bonds are said to be _____ and have a _____ melting point than fatty acids with no double bonds.

8 Amino acids are _____, acting as both an acid and a base. Through condensation of two amino acids they form _____.

9 Amino acids can join by peptide linkages forming _____. New amino acids are always added to the _____ end of the amino acid.

10 The _____ structure of a protein is the sequence of its amino acids. Further structures are controlled by the formation of weak _____ bonds and _____ bridges between amino acids.

If you got them all right, skip to page 21

15

Organic molecules

Improve your knowledge

1 Water is a **dipolar** molecule with two charged ends, which attract ions and other charged molecules forming weak **hydrogen bonds**. This makes water an excellent solvent as water molecules form shells around the solute molecules, e.g. Na^+. It also allows **cohesion** (water molecules cling together) and **adhesion** (water molecules stick to other polar molecules).

Bond angle gives water molecule a dipolar nature

Water also has the following properties:

- **High heat of vaporization** A large amount of heat is needed to turn a given amount of liquid water into water vapour.
- **High specific heat capacity** A large amount of heat is needed to raise the temperature of 1 gram of water by 1°C.

The properties of water are vital to living organisms. It acts as a solvent, metabolite and prevents rapid temperature change

2 Carbohydrates all contain carbon, hydrogen and oxygen, with the general formula $C_x(H_2O)_y$. Their three main functions are:

- As a main respiratory substrate, e.g. glucose.
- As food storage compounds, e.g. starch.
- As a structural substrate, e.g. cellulose.

Monosaccharides are the basic units (monomers) which make up all other carbohydrates. They are soluble in water, sweet and consist of small molecules, characterised by their number of carbon atoms:

- **Triose sugars** Three carbon atoms, e.g. phosphoglyceraldehyde formed in the light-independent stage of photosynthesis.
- **Pentose sugars** Five carbon atoms, e.g. ribose in RNA and deoxyribose found in DNA.

- **Hexose sugars** Six carbon atoms, e.g. fructose found in nectar and fruits to attract insects, and glucose, the main respiratory substrate. Glucose occurs in two forms:

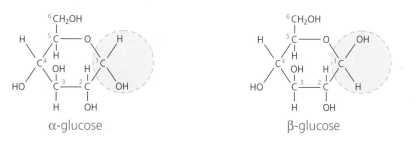

Notice the difference between α- and β-glucose in the position of the OH and H on carbon atom 1

α-glucose β-glucose

3 **Disaccharides** are composed of two monosaccharides joined in a **condensation reaction** (involving the loss of water) forming a **glycosidic** linkage, e.g. α-glucose + β-fructose = sucrose.

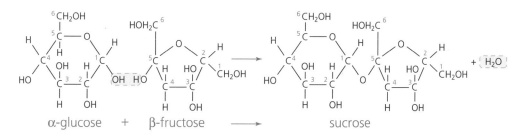

α-glucose + β-fructose ⟶ sucrose

Three common disaccharides are:

- **Sucrose** Transport carbohydrate in plant phloem tissue.
- **Maltose** (α-glucose + α-glucose) Found in germinating seeds.
- **Lactose** (α-glucose + galactose) Found in milk, providing the energy source for suckling mammals.

4 All monosaccharides, along with maltose and lactose, have a carbonyl group (C=O) available to donate electrons, so they act as reducing agents (reducing sugars). This means that these sugars reduce blue Benedict's solution (Cu^{2+}) to an orange-red precipitate (Cu^+).

5 **Polysaccharides** are formed by multiple condensation reactions of monosaccharides. They are not sweet or soluble in water and not crystalline. Their main function in plants and animals is to provide structure and storage because they are insoluble and compact.

Storage molecules are insoluble so they do not affect the movement of water across the cell membrane

Starch	Glycogen	Cellulose
Polymer of α-glucose found in two forms: • **Amylose** – straight unbranched helix • **Amylopectin** – branched chain (branching every 24th glucose monomer)	Formed of branched chains of α-glucose (branching every 10th glucose monomer)	Polymer of β-glucose in an unbranched chain. OH groups stick outwards forming H-bonds with neighbouring chains giving a high tensile strength and permeability
Storage carbohydrate in plants (found in chloroplasts and root cortex parenchyma cells)	Storage carbohydrate in animals, found in muscles and liver, readily hydrolysed to glucose	Form plant cell wall, combined with a gel-like organic matrix

6 **Lipids** contain carbon, hydrogen and oxygen (although lower amounts of oxygen than carbohydrates) and are compact and insoluble. **Non-essential fatty acids** can be synthesised by the body from carbohydrate and protein metabolism. **Essential fatty acids** cannot be synthesised and must be obtained from the diet, e.g. vegetables and seed oils.

Lipids have a variety of functions:

* **Storage** of metabolic fuels in adipose tissue of animals and oils in plants, e.g. castor oil seeds and fruits (lipids contain double the amount of energy of carbohydrates).
* **Heat insulation** in hibernating and marine mammals, e.g. whales.
* **Buoyancy** and **waterproofing** in aquatic animals, e.g. geese.

All lipids are composed of glycerol and fatty acids.

Glycerol Molecular arrangement is the same in all lipids.

Fatty acids The nature of a lipid depends on the particular fatty acids it contains.

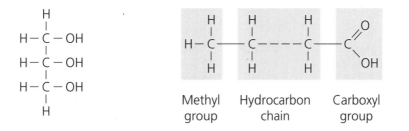

Methyl group	Hydrocarbon chain	Carboxyl group

The general formula of lipids is $C_nH_{2n}O_2$.

7 **Unsaturated fatty acids** form oils, e.g. oleic acid in plants, as they have double bonds between carbon atoms which lower the melting point. **Saturated fatty acids** form fats, e.g. stearic acid in animals, as they have no double bonds, raising the melting points and forming animal fats.

Unsaturated fatty acids do not contain the full number of H atoms

Lipid formation occurs when three fatty acids combine with one glycerol to form **triglycerides** by a condensation reaction, with the loss of three water molecules.

The carboxyl group of each fatty acid bonds to an OH group of glycerol (ester bond)

8 Proteins contain carbon, hydrogen and oxygen, along with nitrogen and often sulphur and phosphate. They are polymers of 20 types of amino acids.

Amino acids have the following characteristics:

- **Amphoteric** They act as both an acid and a base depending on surrounding pH. This allows them to act as a buffer to pH change.
- **Zwitterions** They possess both a negative and positive charge.
- **Form dipeptides** through condensation of two amino acids.

You must know the structure of an amino acid

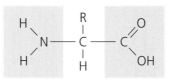

Amino group (basic)	Carboxyl group (acidic)

9 Polypeptides are long chains of amino acids joined by **peptide linkages** following condensation reactions, with new amino acids always added to the amino end (NH_2). The resultant proteins have the following functions:

- Intracellular structure and framework of cells, e.g. keratins.
- Cellular function, e.g. muscle fibres.
- Enzymes.

10 Protein structure can be divided into four levels:

- **Primary structure** This is the sequence of amino acids determined by genetic code of DNA. This determines further structures through controlling the formation of H-bonds and disulphide bridges between amino acid (peptide) chains.
- **Secondary structure** This is the way in which the primary structure folds. In structural proteins this forms fibres, e.g. keratin found in hair and feathers is a two stranded α-helix, whereas keratin found in silk is a β-pleated sheet. These are both formed by hydrogen bonding, but the different structure gives different functional characteristics.
- **Tertiary structure** This is the further folding onto the secondary structure, due to H-bonding and disulphide bridges, forming either 3-D globular proteins, e.g. enzymes and myoglobin, or fibrous structural proteins, e.g. collagen (in bones and tendons).
- **Quaternary structure** Some proteins are made of several polypeptide chains locked together, e.g. haemoglobin consists of two α- and two β-polypeptide chains around an iron-containing haem group.

Now learn how to use your knowledge

Organic molecules

Use your knowledge

1 (a) Define the term 'essential fatty acid'.
  (b) Give two ways in which phospholipids are different from triglycerides.
  (c) State three elements present in fatty acids.

2 For each of the following statements state whether it refers to monosaccharides, fatty acids or amino acids.

  (a) Always contain nitrogen.
  (b) Produced in the complete hydrolysis of cellulose.
  (c) Bonded together by glycosidic links.
  (d) Insoluble in water.

3 The diagram below shows two amino acids:

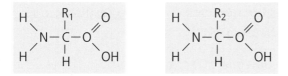

  (a) These two amino acids may be linked to form a new molecule.
    (i) State the type of reaction that links amino acids and name the molecule that is lost.
    (ii) Name the bond formed by this reaction.
    (iii) What is the name of the new molecule formed?
  (b) Name two elements, other than carbon, hydrogen and oxygen, that may be present in R_1 and R_2.
  (c) With reference to the answer you have given in part (b), explain how the presence of different elements at R_1 and R_2 is important in controlling the structure and function of proteins.

Answers on page 91

Protein synthesis and nuclear division

20 minutes

Test your knowledge

1 Transcription involves the production of a _____ of genetic code and _____ is the conversion of the template into a polypeptide.

2 Transcription produces _____ controlled by the enzyme DNA-dependent RNA polymerase, which _____ the DNA α-helix.

3 Translation requires two nucleic acid molecules: _____ molecules to carry amino acids, and a _____ to translate the mRNA strand.

4 Within the ribosome, _____ tRNA molecules bind to the mRNA allowing amino acids to form a _____ bond.

5 Mitosis is involved in _____ reproduction, since the two resultant daughter nuclei produced are genetically _____ to the parent nuclei.

6 During _____ the chromosomes become attached to spindle fibres. They line up along the _____ of the cell.

7 Meiosis produces _____ nuclei from a diploid parental nucleus. The cells produced form _____ responsible for sexual reproduction.

8 In prophase I, homologous chromosomes pair up to form _____. These are linked at points along their length called _____.

9 Anaphase II is the separation of the daughter _____ by pulling apart centromeres, producing _____ haploid cells.

10 Meiosis increases genetic variation through _____ and the _____ arrangement of chromosomes during metaphase I.

Answers

10 chiasmata, independent
7 haploid, gametes   8 bivalents, chiasmata   9 chromatids, four
4 two, peptide   5 asexual, identical   6 metaphase, equator
1 template, translation   2 mRNA, unzips   3 tRNA, rRNA

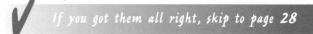

If you got them all right, skip to page 28

Protein synthesis and nuclear division

20 minutes

Improve your knowledge

1 **Protein synthesis** is the production of polypeptide chains based on the genetic code in DNA. It occurs in two stages in all cells that contain a nucleus. **Transcription** is the production of a template of the relevant genetic code and **translation** converts the template into a polypeptide.

Transcription occurs in the nucleus. It is the production of a chain of the nucleic acid **mRNA** from DNA under the control of the enzyme **DNA-dependent RNA polymerase with ligase activity**.

DNA and RNA are both nucleic acids composed of nucleotide units

2 RNA polymerase recognises and binds at the start of the relevant DNA section (**coding strand**). The polymerase now travels along the coding strand, breaking the H-bonds and **unwinding the DNA α-helix**. This exposes bases of the DNA nucleotides, and RNA nucleotides with complementary bases come in to form the single mRNA strand.

mRNA is complementary to the DNA strand

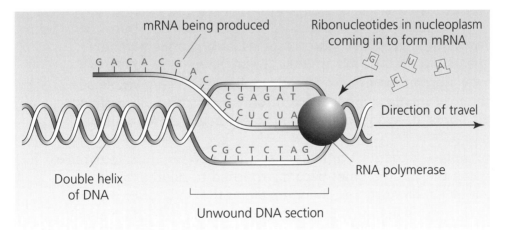

mRNA being produced

Ribonucleotides in nucleoplasm coming in to form mRNA

G A C A C G
A C

C G A G A T
G
C U C U A

Direction of travel

C G C T C T A G

Double helix of DNA

RNA polymerase

Unwound DNA section

The polymerase detaches when it reaches and recognises the stop codon on the DNA. The DNA rewinds and hydrogen bonds reform. The strand of mRNA now leaves the nucleus via a pore into the cytoplasm.

You must know the base pairing rule

3 **Translation** occurs in the cytoplasm and requires tRNA molecules (to bring in amino acids) and a ribosome (to 'translate' the mRNA):

- **Ribosomes** are composed of 50% protein and 50% rRNA, and are made of two subunits which remain separate until mRNA is present.
- **tRNA** is a single strand of RNA, coiled into a clover-leaf shape.

The two ribosome subunits lock onto the ribosome binding site on the mRNA strand. Each ribosome covers two tRNA binding sites on the mRNA.

tRNA molecules carry base **anticodons** complementary to base codes on the mRNA. The anticodon on the tRNA corresponds to a **specific amino acid** that the tRNA carries.

Each mRNA code brings in a tRNA with a specific amino acid

4 Two tRNA molecules bind to the mRNA within the ribosome and the amino acids carried on the tRNA form a peptide bond together. The first tRNA then disconnects from the mRNA leaving a binding site empty. The ribosome moves along one codon and a third tRNA molecule binds carrying another amino acid to bind to the developing chain. The ribosome continues moving until a **stop codon** is reached.

5 **Mitosis** is the division of a nucleus (nuclear division) following chromosome replication. The two resultant daughter nuclei have the same number of chromosomes as the parent nuclei (**genetically identical**). Mitosis is involved in asexual reproduction, and growth and repair of body tissues.

6
- The first stage is **prophase**. Chromosomes slowly condense and become visible, and the microtubules break down making the cell spherical. Chromosomes have already replicated and are visible as **daughter chromatids** joined at a centromere. Microtubules, called **spindle fibres**, develop from the centrioles (only present in animal cells).
- **Prometaphase** is the start of **metaphase**. The nuclear envelope breaks down and the chromosomes become attached to spindle fibres, at regions called **kinetochores**, lining up along the equator of the cell. By the end of metaphase, all the chromosomes have lined up along the metaphase plate, with the spindles pulling towards the poles.
- **Anaphase** begins when the kinetochores split and the daughter chromatids are pulled to opposite poles of the cell by spindle fibres.

- **Telophase** can be considered as the reverse of prophase. Chromosomes begin to recondense and new nuclear membranes reform around the two sets of chromosomes.

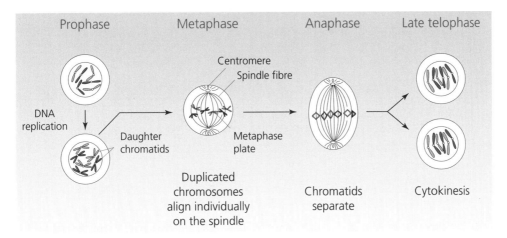

Cytokinesis is cytoplasm division. It is not part of nuclear division but begins when this is complete. Filaments of the proteins actin and myosin contract to produce a **cleavage furrow** at the equator of the cell and continue until two new cells are produced.

Diploid cells contain a complete set of chromosomes

7 **Meiosis** is also nuclear division following replication of chromosomes, but involves two successive divisions. The four resultant daughter nuclei are **haploid** (half as many chromosomes as the diploid parental cells).

Meiosis occurs in **gamete production** in organisms that reproduce sexually. In humans, normal cells contain 46 chromosomes (23 homologous pairs). Gametes contain 23 unpaired chromosomes, so that the fusion of gametes at fertilisation **restores the diploid number**.

8
- In **prophase I**, chromosomes condense and homologous pairs come together, forming **bivalents**. The bivalents are linked at points along their length (at **chiasmata**) and this linkage results in crossing over of genetic material between chromosomes. Chiasmata are random and result in new combinations of genes (**genetic recombinants**).
- **Metaphase I** is identical to metaphase in mitosis, except the chromosomes line up as bivalents on the spindle, via the kinetochore fibres.
- The homologous chromosomes separate during **anaphase I**, with one member of each pair moving to the poles by sliding of kinetochores.

Division 1 separates the homologous pairs of chromosomes

- **Telophase I** is the final stage of the first division and is the reverse of prophase. Cytokinesis occurs, producing two cells containing one member of each homologous pair of chromosomes. Each chromosome is still made up of two daughter chromatids.

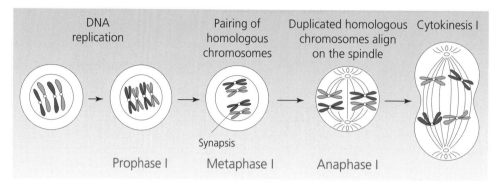

DNA replication — Prophase I
Pairing of homologous chromosomes — Synapsis — Metaphase I
Duplicated homologous chromosomes align on the spindle — Anaphase I
Cytokinesis I

9 The second meiotic division occurs in both cells produced during division I. In **prophase II** chromosomes attach to spindle fibres. The chromosomes then line up individually on the spindle fibre at the equator of the cell in **metaphase II**. **Anaphase II** is the separation of the daughter chromatids by pulling apart centromeres. **Telophase II** produces four haploid cells that are genetically different, following cytokenesis.

Division 2 separates daughter chromatids

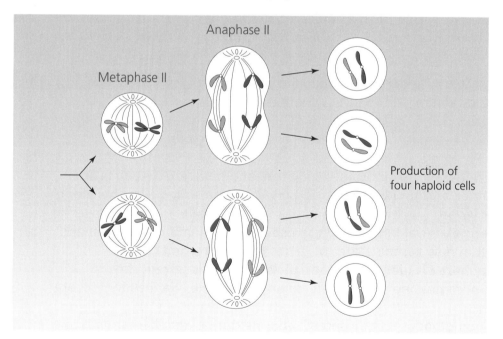

Metaphase II

Anaphase II

Production of four haploid cells

10 Meiosis is significant because it conserves the number of chromosomes during sexual reproduction. If a gamete of an organism had the diploid number of chromosomes, the number of chromosomes would double every generation.

Meiosis also **increases genetic variation** in two ways:

- The number of possible chromosome combinations is very large in a gamete, due to the independence of chromosome arrangement on the spindle at metaphase I.
- Chiasmata causes recombination of paternal and maternal genetic material in the gametes, increasing variation in the population.

Cells spend only 10% of their lifetime in nuclear division, and 90% in **interphase**. During interphase the cell is metabolically active, synthesising new proteins and DNA. The chromosomes are uncondensed forming fine threads. The threads consist of **nucleosomes**, short lengths of DNA wound around a 'bead' of **histone proteins**.

In most organisms, differentiated tissue cells lose the ability to divide. When they die, these cells are replaced by division from undifferentiated stem cells.

Protein synthesis and nuclear division

Use your knowledge

1 The diagram below shows translation occurring on a ribosome:

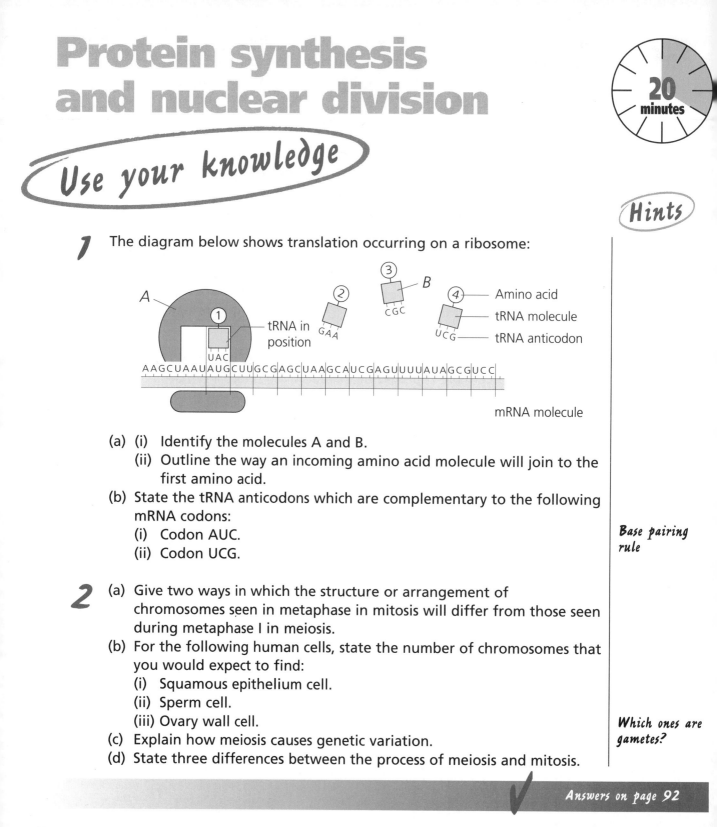

(a) (i) Identify the molecules A and B.
  (ii) Outline the way an incoming amino acid molecule will join to the first amino acid.
(b) State the tRNA anticodons which are complementary to the following mRNA codons:
  (i) Codon AUC.
  (ii) Codon UCG.

2 (a) Give two ways in which the structure or arrangement of chromosomes seen in metaphase in mitosis will differ from those seen during metaphase I in meiosis.
(b) For the following human cells, state the number of chromosomes that you would expect to find:
  (i) Squamous epithelium cell.
  (ii) Sperm cell.
  (iii) Ovary wall cell.
(c) Explain how meiosis causes genetic variation.
(d) State three differences between the process of meiosis and mitosis.

Answers on page 92

Genetics

Test your knowledge

15 minutes

1. Genetics is the study of how _____ are passed between generations. A _____ cross is between two individuals which differ in one characteristic.

2. Mendel identified _____ characteristics which one parent may show and do not appear in the F1 generation, but _____ in the F2 generation.

3. The law of _____ states that an organism's characteristics are determined by pairs of factors (alleles), but only one factor of the pair enters a gamete.

4. Characteristics are controlled by _____ of alleles, which are found at similar loci on _____ chromosomes.

5. The _____ comprises the alleles an individual contains while the _____ concerns its appearance or physiology.

6. Studying _____ crosses, which are crosses between individuals differing in two pairs of characteristics, Mendel made the deduction that alleles for different characteristics are inherited _____.

7. The law of _____ _____ was Mendel's second law. This states that either member of a pair of alleles can combine with either of a second pair during _____ formation.

8. _____ for different characteristics which are carried on _____ pairs of homologous chromosomes behave independently.

If you got them all right, skip to page 34

Genetics

Improve your knowledge

20 minutes

1 Genetics is the study of inheritance, i.e. how characteristics are passed from one generation to another. **Mendelian genetics** are patterns of inheritance worked out by Gregor Mendel cross-breeding garden peas.

2 **Monohybrid crosses** are crosses between two individuals which differ in one characteristic. Mendel used parents which always produced like offspring in terms of characteristics in self-pollination (**pure breeding**). The first (filial) generation is the **F1** and the second is the **F2**.

Identical pure-breeding parents produce identical offspring

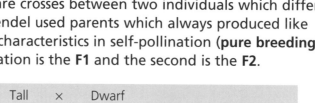

Parents	Tall	×	Dwarf	
F1		Tall		Self-pollination
F2	Tall		Dwarf	
		3 : 1		Monohybrid ratio

Mendel made the following deductions from this cross:

- Parental characteristics are not blended because there are no medium-sized plants.
- Dwarf characteristic is **recessive** and tall is **dominant**, because no dwarf plants occurred in the F1 generation.
- The reappearance of the dwarf plant in F2 means that the F1 plants, which are tall, must contain one factor for tallness and one for dwarfness.

3 This led to **Mendel's first law of segregation**. The characteristics of organisms are determined by pairs of factors. Only one factor of the pair enters a gamete. These factors are now known as **alleles**, which are variations of a gene, i.e. the gene for height can be tall or dwarf.

Gametes are the reproduction or sex cells, e.g. sperm cells

Let T = tallness gene (dominant), t = dwarfness gene (recessive).

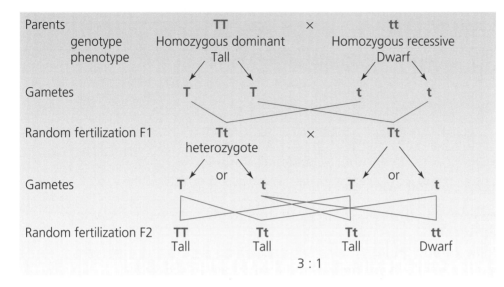

4 **Each individual contains two alleles** controlling a particular characteristic, with one allele inherited from each parent. These pairs of alleles are found at similar positions, or **loci**, on **homologous chromosomes**, i.e. those that pair up during late prophase I of meiosis.

See nuclear division

Because homologous chromosomes are separated at anaphase I during gametogenesis (production of gametes), pairs of alleles are separated. This means that **gametes contain only one member of each pair of alleles**.

Gametes combining in fertilization results in new pairs of alleles

5 A **back cross** or a test cross is used to determine the **genotype** (comprising alleles present) of an individual with a dominant **phenotype** (physical appearance), i.e. a tall individual could be genotype TT or Tt. The dominant phenotype individual is crossed with a recessive individual, whose genotype must be tt.

One dominant allele will cause an individual to show the dominant characteristics

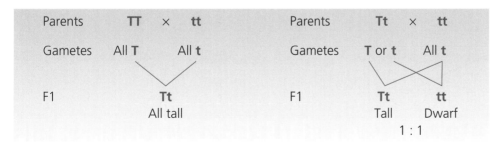

If the F1 generation are all tall the dominant parent was pure breeding (TT). If the F1 are 1:1 for tall:dwarf the dominant parent was Tt.

6 **Dihybrid crosses** are between individuals differing in two pairs of characteristics. The cross below is between parents pure-breeding for both characteristics:

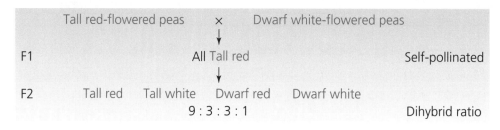

	Tall red-flowered peas	×	Dwarf white-flowered peas	
F1		All Tall red		Self-pollinated
F2	Tall red	Tall white	Dwarf red	Dwarf white	
		9 : 3 : 3 : 1		Dihybrid ratio

Mendel made the following deductions from this cross:

- Since F1 are all tall and red, they are dominant over dwarf and white respectively.
- Two new combinations appear: tall and white, and dwarf and red. Therefore the **alleles for height and colour can be separated**.
- The ratio for tall:dwarf and red:white are both 3:1. Therefore the **alleles are behaving independently**.

7 This led to **Mendel's second law of independent assortment**. Either member of a pair of alleles can combine with either of a second pair when a gamete forms. For example, in the dihybrid cross either of the gametes for height can combine with either for flower colour. Therefore, the F1 **generation has four possible gametes**.

Characteristics are inherited separately if alleles are on separate pairs of chromosomes

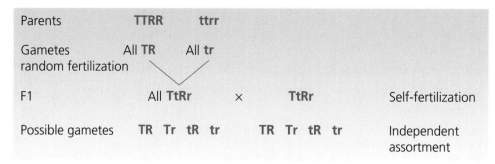

Parents	**TTRR**	**ttrr**	
Gametes random fertilization	All **TR** All **tr**		
F1	All **TtRr** ×	**TtRr**	Self-fertilization
Possible gametes	**TR Tr tR tr**	**TR Tr tR tr**	Independent assortment

A **punnett square** is used to show possible offspring from the random fertilization between gametes:

Gametes	TR	Tr	tR	tr	Female
TR	TTRR	TTRr	TrRR	TrRr	
Tr	TTRr	TTrr	TrRR	Ttrr	
tR	TtRR	TtRr	ttRR	ttRr	
tr	TtRr	Ttrr	ttRr	ttrr	
Male					

F2 phenotypes: nine tall red, three dwarf red, three tall white, one dwarf white.

8 Independent assortment occurs because **alleles for different characteristics behave independently**, since they are carried on separate pairs of homologous chromosomes. The homologous pairs line up independently on the equator during metaphase I of meiosis. It is a matter of chance which combination of alleles move to a pole together and go into a gamete.

A back cross or a test cross is used to determine the genotype of an individual with dominant characteristics. For example, a tall red plant can be TTRR, TtRR, TTRr or TtRr.

Parents TTRR × ttrr   TtRR × ttrr
Gametes TR   tr   TR tR   tr

TtRr   TtRr   ttRr
All tall red   Tall red   Tall white
Therefore original plant homozygous   Therefore original heterozygous for tall
dominant for tallness and redness   and homozygous red

Parents TTRr × ttrr   TtRr × ttrr
Gametes TR Tr   tr   TR Tr tR tr   tr

TtRr   Ttrr   TtRr   Ttrr   ttRr   ttrr
Tall red   Tall white   Tall red Tall white Dwarf red Dwarf white
Therefore original homozygous tall   Therefore original plant heterozygous
and heterozygous red   for height and colour

Genetics

Use your knowledge

20 minutes

Hints

1 The gene for petal colour in a flowering plant has two alleles; white (W) and pink (w). Leaf shape is controlled by a single pair of alleles, round (R) and oval (r), carried on a separate homologous chromosome.

(a) Distinguish between the term gene and allele.

(b) State the phenotype of the following genotypes:
   (i) WWrr.
   (ii) WwRr.

(c) State the possible genotypes of the following plants:
   (i) White flower with round leaves.
   (ii) White flower oval leaves.

2 Peach trees show variation in their height (tall or dwarf) and leaf shape (straight or cut). The following crosses were carried out.

- **Cross 1** Straight-leaved, tall plants were crossed with cut-leaved, dwarf plants. The F1 generation were all straight-leaved, tall plants.
- **Cross 2** The F1 generation were self-fertilized, giving: 899 straight-leaved tall plants, 301 straight-leaved dwarf plants, 290 cut-leaved tall plants, 99 cut-leaved dwarf plants.

(a) State suitable letter symbols for the following alleles:
   Straight–leaved; cut-leaved; tall; dwarf.

(b) Complete the following genetic diagram for the F1 self-fertilization:

F1 phenotype	Straight-leaved tall	x	Straight-leaved tall
F1 genotype	_____		_____
Gametes	_____		_____

Remember capital letters for dominant

(c) Suggest a reason why the observed ratio in the F2 generation is not a precise 9 : 3 : 3 : 1.

Answers on page 92

Mutation and evolution

Test your knowledge

1 Frameshift mutation can involve either _____, where a single base is removed, or insertion, where a new _____ is inserted into the sequence. Whereas _____ mutations involve either substitution or inversion.

2 _____ mutations are changes in chromosome structure during synapsis of prophase I. _____ is where a section of one chromosome becomes attached to another.

3 Evolutionary changes in the characteristics of a population over many generations result in _____. These changes are due to _____ in a population's characteristics due to mutation, independent assortment of chromosomes, chiasmata, and _____ fertilisation.

4 Variation can be either _____, e.g. eye colour, or continuous, e.g. height. _____ selection acts on variations, passing on beneficial characteristics to the next generation.

5 _____ selection removes extremes of a characteristic from a population, whereas disruptive selection removes _____ phenotypes, favouring extremes.

6 Speciation mainly occurs in small isolated populations called _____. The mechanisms of speciation are either by _____ or through isolation, through either allopatric or _____ speciation.

7 _____ speciation occurs through the spatial separation of populations, most commonly through _____ isolation, where populations become separated due to a geographical feature. This prevents _____ exchange between demes.

8 Speciation due to _____ isolation is known as sympatric speciation. For example, _____ isolation occurs when courtship or mating techniques fail to attract or stimulate a potential partner.

Answers

1 deletion, base, non-frameshift 2 chromosome, translocation 3 speciation, variations, random 4 discontinuous, natural 5 stabilising, intermediate 6 demes, polyploidy, sympatric 7 allopatric, geographical, gene 8 reproductive, behavioural

✓ *If you got them all right, skip to page 40*

Mutation and evolution

Improve your knowledge

1 Morphological and functional changes during evolution are produced by mutations (changes in DNA) occurring during DNA replication.

Point **mutations** are changes in the sequence of bases in a gene. There are two main types: frameshift and non-frameshift.

In **frameshift** mutations the whole sequence (reading frame) is altered, i.e. every single codon (three bases coding for an amino acid) after the mutation is changed, forming a nonsense polypeptide. This can be lethal for a cell, if the mutation occurs prior to mitosis, or fatal to the resultant organism if prior to meiotic division.

The closer to the beginning of the gene, the more severe the effect of mutation

There are two types:

- **Deletion**, where a single base is removed.
- **Insertion**, where a new base is inserted into the sequence.

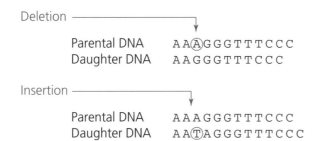

Deletion ————————→

| Parental DNA | A A Ⓐ G G G T T T C C C |
| Daughter DNA | A A G G G T T T C C C |

Insertion ————————→

| Parental DNA | A A A G G G T T T C C C |
| Daughter DNA | A A Ⓣ A G G G T T T C C C |

In **non-frameshift** mutation the whole reading frame is not altered, i.e. apart from the mutated codon, all other codons remain intact, so only one amino acid is different in the resultant polypeptide. This affects the structure and functioning of the protein only if the amino acid is either involved in the folding of the protein or is located within an enzyme-active site.

There are two types:

- **Substitution**, where a wrong base is substituted for the correct base.
- **Inversion**, where two bases in the chain swap places.

Substitution ─────────────┐
                          ▼
Parental DNA   A A A G G G T T T C C C
Daughter DNA   A A Ⓣ G G G T T T C C C

Inversion ─────────────┐
                       ▼
Parental DNA   A A A G G G T T T C C C
Daughter DNA   A A ⓖⓐ G G T T T C C C

2 **Chromosome mutations** involve changes in the chromosome structure during synapsis of prophase I, when chromatids break and rejoin. This changes the gene sequence. Changes can be:

- **Deletion** Sections of DNA are lost (often fatal).
- **Inversion** Sections within the chromosome are reshuffled.
- **Duplication** A section of DNA is doubled.
- **Translocation** A section of one chromosome becomes attached onto another (this can lead to Down's Syndrome).

3 Evolution is the cumulative change in the characteristics of a population over many generations resulting in **speciation**, the production of two or more species from one original species.

These changes are due to variations in a population's characteristics produced by:

Remember the ways that variation occurs during meiosis

- **Mutation** (point or chromosomal).
- **Independent assortment** of chromosomes during metaphase.
- **Chiasmata**.
- **Random combination** of gametes.

4 There are two types of variation:

- **Discontinuous** Distinct types of phenotype in a population, controlled by a single gene, e.g. eye colour.
- **Continuous** Range of phenotypes, controlled by several genes and the environment, changing over generations, e.g. height.

Variation means organisms have differing abilities to survive. If it gives a favourable characteristic, e.g. improved ability to find food, the organism will be more successful in reproducing, so this characteristic is passed to the next generation. This results in **natural selection**.

5 Natural selection can affect continuous variation in three ways:

- **Stabilising selection** Selection against extremes of a characteristic. For example, the optimum weight for human babies is 3.6 kg. Babies born above or below this weight show increased mortality levels.
- **Directional selection** Changes in a phenotype in one particular direction, e.g. the fastest cheetah will be most successful at catching prey, so increasing speed will be selected for within the species.
- **Disruptive selection** Selection against intermediate phenotypes, favouring extremes. It is rare but important for speciation.

6 Speciation only occurs in small isolated populations (**demes**) of a species, which can only interbreed. This is because they have similar genotypes and experience the same environmental conditions affecting their phenotype. The mechanisms of speciation are:

- Polyploidy.
- Isolation (allopatric and sympatric speciation).

A polyploid is an individual whose chromosome number is a multiple of the haploid number of chromosomes and therefore can interbreed only with like individuals. An example is the fertile hybrid cord grass *Spartina angelica*.

Polyploidy is rare in animals because the sex chromosome system breaks down

38

7 Speciation through the spatial separation of populations is known as **allopatric speciation**. The most common way populations become isolated is **geographical isolation**, whereby populations become separated due to geographical features and events, such as:

- Mountain ranges.
- Occurrence of deserts across productive land (desertification).
- Plate tectonics (continents moving apart).

These barriers prevent gene exchange between demes, with mating occurring only within the isolated demes. This results in genetic differences (divergence) arising due to the selection pressure of the different environments and eventually the formation of a new species, unable to interbreed with members of different demes.

8 **Sympatric speciation** occurs in demes found in the same geographical area. It is speciation due to reproductive isolation, i.e. individuals are unable to produce sterile offspring. This involves various mechanisms:

- **Ecological isolation** Species live within different habitats in the same area and rarely meet, e.g. insect parasite populations specific to one plant host.
- **Seasonal isolation** Populations produce gametes at different times of the year, e.g. rainbow trout breed in the spring but brown trout breed in autumn.
- **Behavioural isolation** Courtship or mating techniques, chemical, auditory or visual signals, fail to attract or stimulate a potential partner in a subspecies, e.g. female silkworms release specific chemicals (pheromones) that are recognised only by some males.
- **Mechanical isolation** Structural differences between genitalia prevent fertilisation of gametes.
- **Gamete isolation** Physiological incompatibility of gametes prevents development.
- **Hybrid inviability** A hybrid zygote forms but the individual either fails to achieve sexual maturity or is infertile.

Mutation and evolution

Use your knowledge

Hints

1 Define the following terms and state an example of each:

(a) Chromosomal mutation.
(b) Point mutation.
(c) Codon.
(d) Genotype.

2 Closely-related species of the dog family are found distributed around the earth, all of which are said to have evolved from a common ancestor. For example, dingo are found throughout Australia, coyote along the west coast of North America and jackal in southern Africa.

(a) State the mechanisms of isolation most likely to have produced this variety of species.
(b) Explain how these distinct species may have evolved from a common ancestor.

Think of natural selection in different environments

3 (a) Define the following terms:
   (i) Chromosome variation.
   (ii) Gene variation.
(b) Explain how gene variation may operate in evolution.
(c) (i) What is the difference between allopatric and sympatric speciation?
   (ii) State two examples of sympatric speciation.

Answers on page 92

40

Nutrition

Test your knowledge

15 minutes

1 Photosynthesis is an example of _____ nutrition involving the reduction of CO_2 to organic carbohydrate using _____ energy.

2 _____ are the organs of photosynthesis. They contain _____ mesophyll cells which are closely packed and contain large number of chloroplasts.

3 Photosynthetic pigments are found on the chloroplast's _____ membranes. Pigments are _____ which absorbs red and blue-violet light and accessory pigments the _____.

4 The light-dependent reactions convert _____ in sunlight into chemical energy. _____ is the production of ATP and $NADPH_2$, with electrons from PS2 passing to PS1, whereas non-cyclic photophosphorylation only produces _____.

5 In the light-independent reactions ribulose bisphosphate fixes _____. Triose phosphate produced is used to regenerate _____ and assimilate organic molecules.

6 Digestion occurs through _____ and chemical breakdown. _____ digests starch to _____. This is then broken down to glucose by maltase produced in the crypts of Lieberkuhn of the _____.

7 Products of digestion are absorbed across the ileum by _____ _____ and active transport. The ileum is highly adapted for absorption with finger-like projections, called _____ and a good blood supply to maintain the _____ gradient to take up food.

8 The release of pancreatic juices is triggered by the hormones _____ and _____ which are both secreted by the wall of the duodenum.

9 *Rhizopus* fungi break down _____ organic material by _____ digestion and absorb the digested food. This is known as _____ nutrition.

If you got them all right, skip to page 47

Nutrition

Improve your knowledge

20 minutes

1 **Autotrophic** nutrition is the use of light energy or chemical energy to manufacture the sugars, fats and proteins needed for cellular metabolism. An example of this is **photosynthesis**, the conversion of inorganic carbon dioxide to organic carbohydrate using light energy.

Heterotrophic nutrition involves obtaining energy for cellular metabolism by breaking down complex organic molecules and synthesising of organic molecules from simple organic molecules. The organic molecules are obtained by feeding on other living organisms.

2 Plants are adapted to photosynthesise efficiently by maximising absorption of sunlight and maintaining a supply of raw materials.

Structure	Function	Adaptation
Leaf	Organ of photosynthesis	• Large, flat surface area (lamina) • Supported by turgor pressure and veins • Supply of water and minerals via veins and midrib.
Palisade mesophyll cells	Main photosynthetic cells	• Large number of chloroplasts in cytoplasm • Closely packed in rows under leaf surface • Large intercellular spaces for diffusion in and out of gases.
Chloroplast	Site of photosynthesis	• Thylakoid membranes and grana increasing internal surface area – site of pigment molecules • Stroma (gel-like matrix) containing enzymes for light-independent stage • Contain photosynthetic pigment.

3 Plants use photosynthetic pigments found on the thylakoid membranes of chloroplasts to absorb light energy. The main pigments are **chlorophylls** *a* and *b*, which absorb red and blue–violet light. **Carotenoids** act as accessory pigments absorbing in the blue–violet wavelength range that chlorophylls do not effectively absorb and protect chlorophyll from excess light.

These photosynthetic pigments occur grouped together as two forms: **photosystem 1** (PS1) and **photosystem 2** (PS2). The collection of pigments grouped together maximises efficiency of light entrapment.

4 Photosynthesis occurs in two stages. First, the **light-dependent reactions** convert solar energy in sunlight into chemical energy, as ATP and reduced NADP (NADPH$_2$).

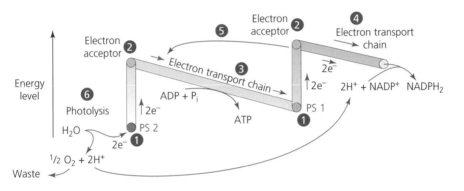

1 Absorption of visible light energy by photosynthetic pigments in PS1 and PS2 causes electrons to achieve an **excited state**.

2 These electrons are picked up at high energy levels by electron acceptors.

3 Electrons from PS2 pass through a series of electron carriers of progressively lower energy levels. The energy released is used to form ATP – **photophosphorylation**. The electrons are finally acquired by positively charged and unstable PS1.

4 Electrons from PS1 can be passed along a series of electron carriers and with protons from H$_2$O reduce NADP to NADPH$_2$ – **non-cyclic photophosphorylation**.

5 The electrons can alternatively be recycled to PS1 and produce ATP – **cyclic photophosphorylation**.

6 PS2 obtains replacement electrons from photolysis of H$_2$O (splitting H$_2$O using light energy). This also releases protons (H$^+$) and O$_2$.

Solar energy is transferred to the electrons exciting them to higher energy levels

PS1 and PS2 must replace excited state electrons to become stable

5 Second, the light-independent reaction fixes CO_2 producing organic
molecules using the chemical energy from the light-dependent reaction.

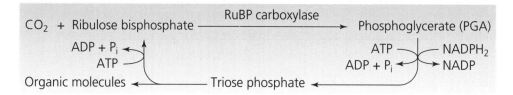

- Five-carbon RuBP accepts CO_2, catalysed by the enzyme RuBP
  carboxylase.
- The unstable six-carbon molecule breaks down to two molecules of
  three-carbon PGA.
- PGA is reduced to triose phosphate using the $NADPH_2$ and ATP (from
  the light-dependent reaction).
- Triose phosphate is used both to regenerate RuBP (requiring ATP) and
  assimilate organic molecules, mainly glucose and starch.

6 Heterotrophic nutrition in mammals requires a specialised gut to digest
then absorb food.

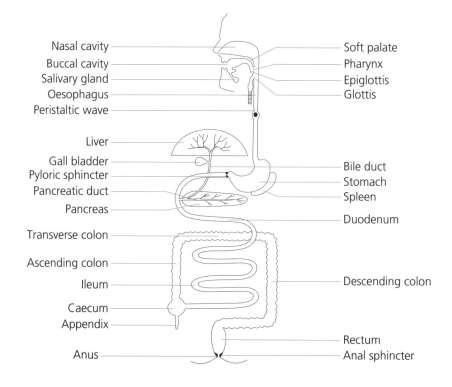

You must be able to label the mammalian gut

Digestion occurs through **mechanical breakdown** (chewing food and contractions of the gut wall), which increases surface area for **chemical breakdown** by enzymes.

Food is digested into its soluble constituents to allow absorption into the blood

Enzyme	Site of production	Digestive action Substrate	Product(s)
Salivary amylase	Parotid salivary gland (mouth)	Starch	Maltose
Endopeptidases e.g. pepsin	Gastric pits in stomach (pepsin secreted as pepsinogen, activated by HCL)	Proteins	Polypeptide
Exopeptidases e.g. trypsin	Pancreas (acting in duodenum)	Polypeptides	Amino acids
Pancreatic amylase	Exocrine region of pancreas (acting in duodenum)	Starch	Maltose
Pancreatic lipase	Pancreas	Lipids	Fatty acids and glycerol
Lactase	Cells in crypts of Lieberkuhn of duodenum (small intestine)	Lactose	Glucose and galactose
Maltase		Maltose	Glucose
Sucrase		Sucrose	Glucose and fructose

Bile produced by the liver is not a digestive enzyme; it **emulsifies** fats (break up into small droplets), increasing surface area for lipase.

Waves of contraction in the smooth-muscle cells of the gut wall (mucosa and muscularis externa), called **peristalsis,** constrict progressive regions of the gut squeezing food along.

7 When digestion is complete the products are absorbed across the ileum via **facilitated diffusion** and **active transport**.

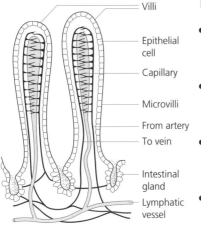

Villi

Epithelial cell

Capillary

Microvilli

From artery

To vein

Intestinal gland

Lymphatic vessel

The ileum is highly adapted for absorption:

- Single layer of cells of the epithelium provides a short distance for uptake of digested food.
- Folds in the wall increase surface area. These folds have finger-like projections (**villi**), the cells of which have microvilli.
- Good blood supply maintains the concentration gradient rapidly removing uptaken food.
- In adult humans the small intestine is 6m long.

Active transport allows digested food to be absorbed against a concentration gradient

8 Control of digestive juice secretion is by **nerve cells and hormones**. Nervous reflexes cause secretion of saliva by the salivary glands, due to sight, smell or taste, and the initial release of gastric juices.

Hormones are released into the blood and travel to where they have an effect

Hormone and source	Action	Stimulus
Gastrin (endocrine cells in stomach wall)	Continual release of gastric juices	Arrival of food in the stomach
Secretin (wall of the duodenum)	Release of pancreatic juices	Arrival of chyme (from stomach) in duodenum
Cholecystokinin (wall of the duodenum)	Release of pancreatic juices and release of bile from gall bladder	Arrival of chyme (from stomach) in duodenum

9 **Saprophytic nutrition** is a type of heterotrophic nutrition whereby organisms, e.g. *Rhizopus* fungi, obtain their organic molecules from dead or decaying plant and animal material, by releasing enzymes externally (**extracellular digestion**) and absorbing the digested food.

Parasitic nutrition involves obtaining organic molecules by living on (ectoparasites) or in (endoparasites) a host organism, causing some degree of harm. For example, *Taenia*'s (tape worm) primary host is humans, where it lives in the intestine absorbing the pre-digested food.

Now learn how to use your knowledge

Nutrition

Use your knowledge

1 (a) State where in the chloroplast the light-dependent reaction takes
       place.
   (b) In the light-dependent reaction molecules are broken down,
       producing oxygen, electrons and hydrogen ions.
       (i)   What is the name given to the process of breaking water down?
       (ii) What is the fate of the electrons produced?
       (iii) What happens to the hydrogen ions?

2 Three plants of the same species, size and age (A, B and C) were placed in
   varying conditions of light and temperature.

   (a) In the first experiment all the plants were kept in the same
       conditions, except: plant A received white light; plant B received red
       light; and plant C received green light.
       (i) State and explain which plant will have the fastest rate of
            photosynthesis.
       (ii) State and explain which plant will have the slowest rate of
            photosynthesis.
   (b) State two factors other than light which will affect the rate of
       photosynthesis.

3 (a) Identify the region of human digestive system where the following
       occur:
       (i)   Production of hydrochloric acid.
       (ii) Emulsification of fats.
       (iii) Facilitated diffusion of glucose molecules.
   (b) Describe three ways in which the structure of the ileum is adapted to
       the function it performs.
   (c) Explain the role of the smooth muscle in the wall of the ileum.

*Which
wavelengths
of light are
used in
photosynthesis?*

Answers on page 92

Respiration and gaseous exchange

15 minutes

Test your knowledge

1 Respiratory organs have specific characteristics to increase their effectiveness. These include being _____ , so gases can dissolve, having a _____ surface to minimise diffusion distance and a large surface area.

2 _____ is the movement of air in and out of the lungs. This occurs by changing the volume of the _____ . During _____ , the external intercostal muscles contract, moving the rib cage up and out, and the diaphragm contracts.

3 Increased cellular respiration _____ blood CO_2 concentration and _____ blood pH. This is detected by chemoreceptors. The _____ _____ sends nervous impulses to intercostal muscles and the diaphragm to speed up the rate of inspiration and expiration.

4 _____ are a series of hollow tubules providing the respiratory system of insects. Contraction of thoracic and abdominal muscles flattens and _____ the volume of the body. This _____ pressure forces air _____ _____ the tracheae system. Relaxation of the muscles brings air in.

5 When insects are active, respiring cells produce _____ _____ , which draws the fluid out of the tracheoles via _____ . This allows a _____ volume of air to penetrate and delivers O_2 faster to the active cells.

6 In fish, _____ occurs when the mouth closes and floor of the mouth cavity raises. This reduces volume and increases pressure forcing water out through the _____ . This is known as the _____ pressure pump.

If you got them all right, skip to page 53

Respiration and gaseous exchange

20 minutes

Improve your knowledge

1 **Respiration** is the intake of oxygen and output of carbon dioxide between the body and air. Respiratory organs have the following characteristics to increase their effectiveness:

- **Large surface area** To maximise exchange of gases. Lungs contain over 750 million alveoli (tiny sacs) giving a surface area of 80m².
- **Moist surface** Because gases must dissolve before diffusion can occur.
- **Thin surface** To maximise diffusion rate through minimising distance to travel. Alveoli are only one cell thick.
- **Transport mechanism** To carry gases between the respiratory organs and body cells, e.g. mammals use the respiratory pigment haemoglobin in the blood, which has a high affinity for oxygen.
- **Ventilation** To maintain air flow to respiratory organs.

The lungs are the respiratory organs in humans.

Remember, O_2 is needed for cellular respiration – production of ATP

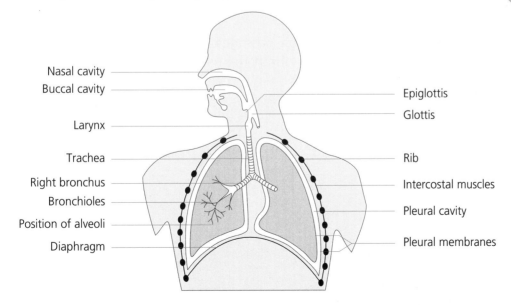

Nasal cavity
Buccal cavity
Larynx
Trachea
Right bronchus
Bronchioles
Position of alveoli
Diaphragm

Epiglottis
Glottis
Rib
Intercostal muscles
Pleural cavity
Pleural membranes

2 Air is moved in and out of the lungs (**ventilation**) by changing the volume of the thorax (body cavity containing the lungs), which changes the air pressure in the lungs.

Inspiration (air moving in)	Expiration (air moving out)
1 External intercostal muscles **contract**, internal ones relax, moving rib cage up and out	1 External intercostal muscles **relax**, internal intercostal muscles contract and ribcage goes down and in under its weight
2 Diaphragm **contracts** (flattens down)	2 Diaphragm **relaxes** (domes upwards)
Increased volume of the thorax **decreases pressure** below atmospheric pressure, so sucking air in	**Decreased volume** of the thorax **increases pressure** above atmospheric pressure, so forcing air out

3 Ventilation rate varies depending on level of activity:

- Exercise causes a demand for more ATP, meaning there is an **increase in cellular respiration**.
- Increased cellular respiration **increases blood CO_2 concentration**, which **decreases pH**.
- **Chemoreceptors** in the pulmonary artery, carotid arteries and medulla oblongata of the brain detect this change in pH.
- **Medulla oblongata** sends nervous impulses to intercostal muscles (via the intercostal nerve) and the diaphragm (via the phrenic nerve) to speed up the rate of inspiration and expiration and the volume of air breathed.
- **Stretch receptors** in the walls of air passages are active during inspiration. Impulses from these inhibit the medulla and prevent over-expansion of the lungs.

CO_2 dissolves in the blood plasma producing carbonic acid

4 The respiratory system of insects consists of **tracheae** (hollow tubules) linking the **spiracles** (openings along the insect's body) to body cells. Tracheae branch to smaller tracheoles, allowing gaseous exchange throughout the body.

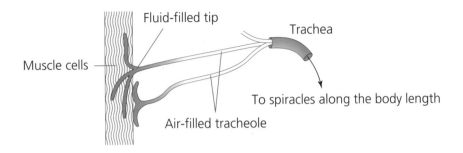

Fluid-filled tip

Trachea

Muscle cells

To spiracles along the body length

Air-filled tracheole

5 Ventilation in insects involves **contraction of thoracic and abdominal muscles**, flattening the body and decreasing volume. The increased pressure forces air out of the tracheae. Relaxation of the muscles increases the volume, reducing pressure and sucking air in through the spiracles.

The end portions of the tracheoles are filled with fluid through which gases diffuse. During increased activity, respiring cells produce **lactic acid**, increasing their concentration and drawing this fluid out of the tracheoles through osmosis. This allows air to penetrate further and deliver O_2 faster to the active cells.

Remember, water will move from a dilute solution to a concentrated solution via osmosis

6 In fish, gills provide the respiratory organs.

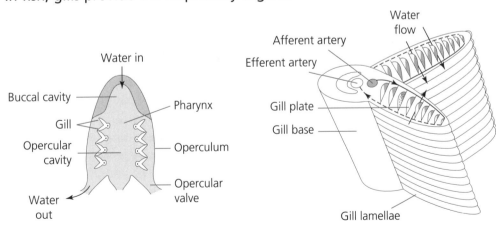

Water in

Buccal cavity

Gill

Opercular cavity

Water out

Pharynx

Operculum

Opercular valve

Water flow

Afferent artery

Efferent artery

Gill plate

Gill base

Gill lamellae

Water will flow over the gills, allowing gaseous exchange when water pressure in the opercular cavity is lower than in the mouth cavity.

Inspiration (opercular suction pump)	Expiration (buccal pressure pump)
• Mouth opens and floor of buccal cavity lowers • Operculum bulges outwards	• Mouth closes and floor of the mouth cavity raises • Muscular contraction force the operculum inwards
Increased volume reduces pressure, sucking water in through the mouth and then over gills	Reduced volume increases pressure forcing water out through operculum

As much O_2 as possible is transferred from the water to the blood in the gill lamellae because the blood flows in the opposite direction to the water flow. This is known as **counterflow**.

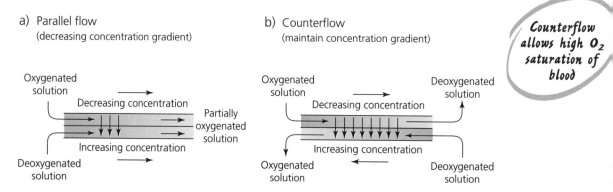

a) Parallel flow
(decreasing concentration gradient)

b) Counterflow
(maintain concentration gradient)

Counterflow allows high O_2 saturation of blood

The diagram shows that with counterflow, as oxygenated water loses O_2 (reduces O_2 concentration) it encounters blood with a decreasing concentration of O_2, so maintaining a concentration gradient.

Now learn how to use your knowledge

Respiration and gaseous exchange

Use your knowledge

15 minutes

Hints

1. (a) Explain why mammals, birds and fish have a complex system for gaseous exchange, whereas very small organisms have no special adaptations for gaseous exchange.
   (b) State three characteristics of a gaseous exchange surface.

Remember surface area to volume ratio

2. (a) Describe how a molecule of carbon dioxide outside the leaves of a flowering plant reaches the spongy mesophyll cells inside a leaf and diffuses into the cytoplasm of the cell.
   (b) State and describe two sites of gaseous exchange in plants, apart from stomata.

3. (a) Describe the route taken by a molecule of oxygen from outside the mouth to reaching the alveoli.
   (b) The bronchioles contain rings of cartilage. What is the importance of this tissue in the functioning of the respiratory tract?

What happens to volume with pressure change?

4. The diagram below is a representation of an air sac in the lungs showing alveoli:

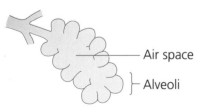

— Air space

Alveoli

(a) How is the structure of the pavement epithelium cells lining the air sac adapted to their function?
(b) Explain why the surface of the alveoli are kept moist.

Answers on page 93

Removal of waste

15 minutes

Test your knowledge

1. In humans, surplus amino acids cannot be stored and so are _____ _____ in the liver. The amino group removed enters the _____ cycle to form urea which is removed from the blood by the _____.

2. The kidneys function both for nitrogenous excretion and _____. Their basic functional units are coiled tubes called _____ where urine forms.

3. The _____ is a dense capillary network found in the Bowman's capsules. The high blood pressure maintained due to the _____ arteriole being narrower than the _____ arteriole results in the process of _____.

4. Selective re-absorption occurs in the proximal convoluted tubule and is the re-absorption of useful substances, e.g. _____. The mechanisms involved include diffusion, _____ _____ and _____ transport. PCT have _____ to increase surface area and mitochondria to manufacture _____ in respiration.

5. The counter-current multiplier occurs in the loop of Henle, where the ascending limb is _____ to water and the descending limb is _____. The concentration of the urine produced is controlled by the hormone _____, which causes the walls of the collecting duct to become _____ to water.

Answers

1 broken down, ornithine, kidneys 2 osmoregulation, nephrons
3 glomerulus, efferent, afferent, ultra filtration 4 glucose/amino acids/sodium ions, facilitated diffusion, active, microvilli, ATP
5 impermeable, permeable, ADH, permeable

If you got them all right, skip to page 58

54

Removal of waste

Improve your knowledge

20 minutes

1 **Excretion** is the removal of the waste products of metabolism and substances in excess from the body, particularly surplus nitrogen-containing compounds, e.g. amino acids, which cannot be stored and are broken down (**deamination**). In humans, deamination occurs in the liver.

- The amino group (NH_2) is removed from the amino acid, forming **ammonia** (NH_3) through the addition of a hydrogen atom.
- Ammonia is highly toxic so is brought into the **ornithine cycle** (cyclical series of reactions), reacting with CO_2 to form less-toxic **urea** which enters the bloodstream and is taken to the **kidney**.
- The remainder of the amino acid is used in cellular respiration.

2 In mammals the kidneys function as organs of **nitrogenous excretion** (filtering urea from the blood) and **osmoregulation** (filtering excess sodium and water). The basic structure is shown below. Each kidney contains a million coiled tubes (nephrons) where urine is formed.

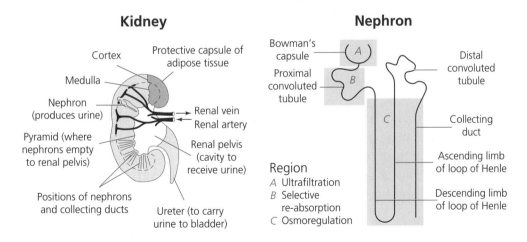

Kidney

Cortex
Medulla
Protective capsule of adipose tissue
Nephron (produces urine)
Pyramid (where nephrons empty to renal pelvis)
Renal vein
Renal artery
Renal pelvis (cavity to receive urine)
Positions of nephrons and collecting ducts
Ureter (to carry urine to bladder)

Nephron

Bowman's capsule
Proximal convoluted tubule
Distal convoluted tubule
Collecting duct
Ascending limb of loop of Henle
Descending limb of loop of Henle

Region
A Ultrafiltration
B Selective re-absorption
C Osmoregulation

3 Bowman's capsules contain dense capillary networks, the **glomeruli**. The blood pressure is high inside these capillaries because:

- Blood in the renal artery arriving at the kidney is at high pressure.
- The **efferent arteriole** removing blood is narrower than the **afferent arteriole** supplying blood.

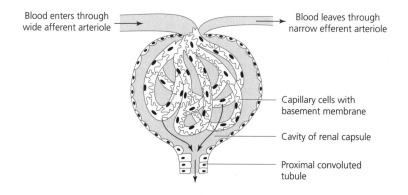

The high pressure forces small soluble molecules, such as **H_2O**, **glucose**, **amino acids**, **sodium chloride and urea**, from the blood, through the pores in the capillary, across the basement membrane surrounding the capillary and into the nephron, forming **glomerular filtrate**. This process is called **ultrafiltration**. Larger molecules, e.g. blood cells and plasma proteins, cannot pass through the filter.

4 Toxic substances in glomerular filtrate, e.g. urea, need to be removed from the body, but many substances present are needed by the body. **Selective re-absorption** occurs in the **proximal convoluted tubule (PCT)**, re-absorbing these substances.

- **Glucose**, **amino acids** and **sodium ions** diffuse from the tubule into the surrounding cells.
- Active transport then pumps these useful molecules out of the cells into intercellular spaces. This maintains a concentration gradient, so diffusion can continue out of the tubule.
- From the intercellular spaces, the molecules diffuse into the capillaries surrounding the PCT, back into the blood.

Active transport requires ATP from respiration

The removal of these soluble substances results in an osmotic gradient between the filtrate and the surrounding cells. This draws water from the tubule via osmosis back into the blood. This allows re-absorption of 80–90% of the water.

PCT is the main site of water re-absorption

Cells of the PCT are adapted for re-absorption with **microvilli** to increase surface area and **many mitochondria** to manufacture ATP.

5 The loop of Henle and the collecting duct form a system known as the **counter-current multiplier** which enables production of urine more concentrated than blood.

Desert mammals, e.g. gerbils, have very long loops of Henle, so produce very concentrated urine to conserve water

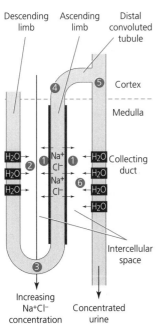

1 Na$^+$ and Cl$^-$ ions are actively pumped into the intercellular space, producing a concentration gradient into the medulla. H$_2$O cannot follow because the **ascending limb is impermeable**.

2 H$_2$O moves out of the descending limb (which is permeable) via osmosis, increasing the filtrate concentration.

3 Filtrate at the base of the loop is very concentrated. When it enters the ascending limb Na$^+$ ions are pumped out.

4 This makes the filtrate at the top of the limb dilute again, allowing more H$_2$O diffusion.

5 The filtrate drains into the collecting duct which flows back through the **concentrated medulla**.

6 The hormone **ADH** causes the walls of the collecting duct to become permeable to water, meaning water osmotically moves out into the surrounding blood capillaries.

Removal of waste

Use your knowledge

1 Experimental techniques are able to analyse the composition of mammalian samples of blood plasma, glomerular filtrate and urine.

Comment on the main difference in the composition of the following:

(a) Glomerular filtrate and urine.
(b) Blood plasma and glomerular filtrate.

2 The graph below shows the concentration of solutes in different regions of a nephron in the human kidney:

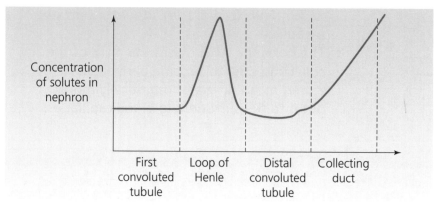

(a) Comment on the graph, saying why the concentration of the solutes increases then decreases as filtrate passes through the loop of Henle.
(b) State the position within the nephron where glucose is re-absorbed into the blood.
(c) Explain why glucose appears in the urine when the concentration of glucose in the blood exceeds 20 mmol dm^{-3}.
(d) Explain why the concentration of solutes does not change in the PCT.
(e) Suggest whether the hormone ADH was present during the production of urine above and give a reason for your answer.

Answers on page 93

Transport in plants and animals

15 minutes

Test your knowledge

1 In plants water and minerals are transported in _____ vessels. In mammals, _____ carry blood under high pressure and veins carry low pressure blood.

2 Three main forces move water through xylem: (1) _____ pressure, (2) _____ : tension, and (3) transpiration.

3 _____ is the movement of mineral and organic compounds through the phloem, possibly according to the _____ hypothesis.

4 Erythrocytes are adapted to carry _____ by their _____ _____ shape and red pigment _____.

5 Haemoglobin takes up O_2 least readily when fully _____ and fully oxygenated haemoglobin dissociates _____ readily.

6 If blood pH decreases, the oxygen-dissociation curve shifts to the _____, because the haemoglobin has a _____ affinity for O_2.

7 Mammals have a _____ circulatory system, as blood passes through the heart twice before being pumped around the body. The two parts of the circulatory system are the _____ and systemic.

8 Following atrial systole, the _____ node is stimulated and the impulse is spread to the apex of the ventricles via conducting tissues in the _____ _____ _____.

9 The cardiac-control centre of the _____ _____ in the hind brain receives information on blood pressure. The sympathetic ANS secretes _____ which speeds up the heart rate. The _____ system secretes acetylcholine which slows down heart rate, acting on the SAN or vagus nerve.

Answers

1 xylem, arteries 2 root, cohesion 3 translocation, mass-flow 4 oxygen, biconcave disc, haemoglobin 5 deoxygenated, most 6 right, reduced 7 double, pulmonary 8 atrioventricular, Bundle of His 9 medulla oblongata, noradrenaline, parasympathetic

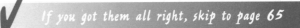

If you got them all right, skip to page 65

Transport in plants and animals

20 minutes

Improve your knowledge

1   Multi-cellular organisms need a transport system to supply every cell with oxygen, water and nutrients because diffusion is too slow to meet demand. Both plants and animals use a system of specialised tubes.

Plants

Tissue	Characteristics	Structure to function
Xylem vessels Carry water and minerals e.g. nitrate	• Made of **xylem vessel elements** and **tracheids** • Have thick, heavily **lignified** secondary cell walls and a hollow central lumen	• Dead, lignified cells provide support to the plant
Phloem Carry sucrose, amino acids, fatty acids, glycerol and hormones	• **Sieve tubes** (cylindrical column of cells) with end walls perforated forming **sieve plates** • **Companion cells** connected to sieve tubes via numerous plasmodesmata	• Companion cells may regulate activity of sieve tube • All cells are living allowing control of transport

Lignification kills cells but makes them very strong and waterproof

60

Mammals

Arteries Carry blood away from the heart	• Lined with **smooth epithelial cells** • **Thick muscular, elastic walls** with the the ability to stretch and recoil • Lumen 100 μm → 3 cm	• Offer low resistance • Can withstand high pressure • Able to maintain continuous pressure	
Capillaries Arranged in beds forming site of exchange to cells	• Lumen 8 μm with wall thickness 2 μm • Pores in capillary wall (with basement membrane) • Numerous, forming large cross sectional area	• Short diffusion distance between cells and blood • Pores allow production of lymph (bathing cells)	
Veins Carry blood back to heart	• Less elastic and muscle fibre than arteries • Pocket-like folds (valves), strengthened by fibrous tissue • Lumen 30 μm → 2.5cm	• Blood flow is assisted by movement of skeletal muscles • Valves prevent backflow of low pressure blood	

Arteries are adapted to resist high pressure, veins to prevent back flow

2 Three main forces move water and minerals up the xylem:

- **Root pressure** Mineral ions (e.g. K^+) are actively pumped into the pericycle cells around the xylem from the adjacent endodermal cells. This moves water into the xylem via osmosis and creates an upward force.
- **Cohesion : tension theory** H_2O molecules are polar and so both attract each other (**cohesion**) and are attracted to the walls of the xylem (**adhesion**). These forces can draw water up even very tall trees.

- **Transpiration** H_2O molecules evaporate from the spongy mesophyll cells in the leaves and diffuse out via the stomata. This lost H_2O causes H_2O from adjacent cells to diffuse in, drawing H_2O from the xylem. The cohesion of H_2O molecules draws water up from the roots.

Various factors will **increase the rate of transpiration**:

- **Increasing light intensity** causing full stomata opening.
- **Increasing temperature** increases kinetic energy of H_2O molecules, causing faster movement.
- **Decreasing humidity** means air can hold more water and so increases the concentration gradient between air and leaves.
- **Increased air movement** removes saturated air from around leaves.
- **Decreased stomata density** prevents overlap of diffusion shells.

Rate of transpiration is dependent on the size of the diffusion gradients

3 **Translocation** is the bi-directional movement of mineral and organic compounds manufactured in the plant, e.g. sucrose in the phloem.

The **mass-flow hypothesis** says that translocation is driven by glucose in the leaves raising their osmotic pressure, causing water to enter. This generates a **turgor pressure**, causing movement along the phloem. At the roots these sugars are removed and water is lost, returning to the leaves.

There are various limitations with the mass-flow hypothesis:
- Sieve plates would resist the turgor pressure.
- Phloem transport is bi-directional.
- It does not explain why sieve tubes and companion cells are metabolically active, nor the role of ATP in translocation.

Phloem tissue is living cells, unlike xylem

4 Mammalian blood is 55% plasma and 45% cells. Plasma consists of water, proteins and inorganic ions, e.g. Na^+. Its functions are:

- **Maintain blood pressure**.
- **Transport** of (i) products of digestion, e.g. amino acids and glucose, (ii) metabolic waste products, e.g. CO_2 and urea, (iii) hormones.

Blood cells comprise **erythrocytes** (red blood cells), **leucocytes** (white blood cells) and **platelets** (cell fragments).

Erythrocytes are adapted to carry oxygen:

- Biconcave disc shape increases surface area for O_2 uptake.
- Cytoplasm is composed of red-pigment haemoglobin, which is specifically adapted for temporary carrying of oxygen.
- Possess no nucleus or organelles and so have more room for haemoglobin.

5 Haemoglobin can either exist in a deoxygenated form or in one of four oxygenated forms:

1. Hb_4 + O_2 ⟷ Hb_4O_2
   DeoxyHb — OxyHb 1

2. Hb_4O_2 + O_2 ⟷ Hb_4O_4
   OxyHb 1 — OxyHb 2

3. Hb_4O_4 + O_2 ⟷ Hb_4O_6
   OxyHb 2 — OxyHb 3

4. Hb_4O_6 + O_2 ⟷ Hb_4O_8
   OxyHb 3 — OxyHb 4

- O_2 is taken up and released least readily by Hb_4 (**deoxyhaemoglobin**).
- O_2 is taken up most readily by Hb_4O_6. Similarly, O_2 dissociates most readily from Hb_4O_8.
- Once haemoglobin has bound O_2 it becomes more reactive with each molecule that binds.
- When haemoglobin has bound its full complement of O_2 it is most unstable.

6 The characteristics of haemoglobin give rise to the sigmoid (s-shaped) curve for the dissociation of O_2 from haemoglobin.

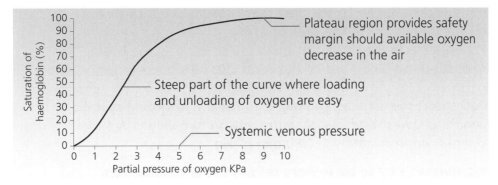

The oxygen-dissociation curve is affected by slight changes in blood pH. These are caused by changes in the levels of carbonic acid, determined by the volume of CO_2 produced through respiration.

- At decreased pH (more **acidic**) the curve shifts to the **right**, as the haemoglobin has a **reduced affinity** for O_2.
- At increased pH (more **alkaline**) the curve shifts to the **left**, as haemoglobin has **increased affinity** for O_2.

So:

↑ Muscular ⟹ ↑ respiration ⟹ ↑ CO_2 ⟹ ↓ pH ∴ Dissociation curve
   activity — produced — shifts to the right.

This is known as the **Bohr effect** and means that more O_2 is released to the cells that require it, i.e. those actively respiring.

Bohr effect means cells with the greatest O_2 demand receive the most

7 Mammals have a **double circulatory system**. Blood passes through the heart twice before being pumped to the tissues of the body. The circulatory system can be divided into two:

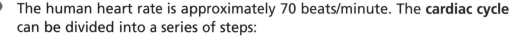

You must be able to label a diagram of a mammalian heart

- **Pulmonary** (including the heart and lungs).
- **Systemic** (including the heart and rest of body). This means the heart has two sides.

8 The human heart rate is approximately 70 beats/minute. The **cardiac cycle** can be divided into a series of steps:

Step 1   Heart beat starts in the **sinoatrial node** (SAN), which is **myogenic**, i.e. generates rhythmic impulses without stimulation from the brain.

Step 2   Impulses spread across the atria, causing **atrial systole**, a uniform contraction emptying the atria into the ventricles. A thin layer of non-conductive tissue prevents the impulse reaching the ventricles.

Step 3   The SAN stimulates the **atrioventricular node** (AVN), which spreads the impulse to the base (apex) of the ventricles via conducting tissues in the **Bundle of His**.

Step 4   **Purkinje fibres** then transmit the impulse back up the heart, squeezing the blood out of the ventricles into the arteries.

9 When blood enters the arteries, the highest pressure is termed **systolic pressure**, and the lowest **diastolic pressure**, giving two values. Normal blood pressure is approximately 120/80 mm of Hg.

Blood pressure is monitored by **sensory cells** in the aorta and right atrium. They transmit impulses via the **vagus** nerve to the **cardiac control centre** of the **Medulla oblongata** in the hind brain. Sensory cells in the **carotid sinus** (in the neck) also relay impulses along the sinus nerve. The cardiac-control centre regulates the SAN through the two parts of the **autonomic nervous system** (ANS):

- **Sympathetic** Secretes noradrenaline which speeds up the heart rate by acting on the SAN and ventricular walls.
- **Parasympathetic** Secretes acetyl choline, which slows down the heart rate by acting on either the SAN or vagus nerve.

Now learn how to use your knowledge

Transport in plants and animals

20 minutes

Use your knowledge

Hints

1 The transport system in flowering plants involves both xylem and phloem tissue.

(a) Name the two types of cells characteristic of phloem tissue.

(b) In an experiment, an aphid is allowed to feed on a plant stem. The experimenter then anaesthetised the aphid and cut off its proboscis (feeding apparatus). The exudate produced was then collected and analysed.

   (i) Suggest which tissue the exudate is being produced from.

   (ii) State two substances which you would expect to find in the exudate.

   (iii) State one substance that will not be present.

Aphids are parasites on plants

2 The diagrams below show cross-sections of human blood vessels (not to scale):

I

II

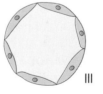

III

(a) Name the three blood vessels labelled I, II and III.

(b) State one way in which each of the vessels are adapted to their function.

(c) Name which of the three vessels has a wall allowing movement of white blood cells in and out.

Answers on page 93

Nerves and muscles

Test your knowledge

1 The peripheral nervous system links receptors and effectors to the _____. It can be divided into the _____ nervous system and the autonomic nervous system.

2 Sensory neurones carry impulses from receptors to _____, where relay neurones link them to _____ neurones.

3 A _____ arc is an involuntary, rapid response to a specific stimulus, in order to _____ the body from damage.

4 A _____ difference is a charge difference across the axon membrane, by movement of Na^+/K^+ ions through membrane proteins. The membranes are impermeable to _____ _____, but permeable to _____ _____.

5 A depolarised membrane is _____ to Na^+ ions. A localised electric current occurs due to charge differences between _____ _____ _____.

6 An impulse arriving at the synaptic knob causes an influx of _____ ions, resulting in movement of vesicles of _____ chemical.

7 _____ post-synaptic potential causes hyperpolarisation. This prevents _____, so an action potential is unlikely to occur.

8 Myofibrils are made of _____ (thin filaments consisting of two helical strands of globular proteins) and _____ (thick filament made of long fibrous proteins with globular heads on a flexible hinge).

9 _____ expose the myosin binding site on actin and ATP binding to myosin causing a change in shape allowing myosin to bind to this site.

Answers

1 CNS, voluntary **2** CNS, motor **3** reflex, protect
4 potential, Na^+ ions, K^+ ions **5** permeable, nodes of Ranvier
6 Ca^{2+}, transmitter **7** inhibitory, depolarisation **8** actin, myosin
9 Ca^{2+} ions

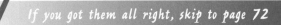

If you got them all right, skip to page 72

50 minutes

1 Nervous co-ordination involves the detection and response to internal changes – **homeostatic regulation**, and the detection and response to external changes for survival, finding food and finding mates. A **stimulus** is any change which provokes a response. **Receptors** are specialised cells that detect a stimulus. **Effectors** are organs that bring about a response – muscles and glands.

The nervous system can be divided into two areas:

- **Central nervous system** (CNS) – brain and spinal cord.
- **Peripheral nervous system** – nerves linking receptors and effectors to CNS.

These can be further divided into:

- **Voluntary nervous system** – conscious control and perception.
- **Autonomic nervous system** – unconscious control (divided into sympathetic system and parasympathetic system).

2 Cell components of the nervous system.

Sensory neurones – Carry impulses from receptors to CNS.

You must be able to label the different types of neurones

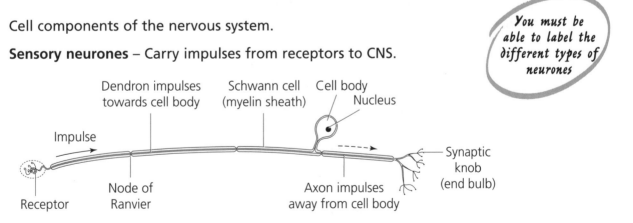

Dendron impulses towards cell body — Schwann cell (myelin sheath) — Cell body — Nucleus

Impulse

Receptor — Node of Ranvier — Axon impulses away from cell body — Synaptic knob (end bulb)

Relay/association neurones – Link sensory and motor neurones. They make up the majority of brain and spinal cord (grey matter).

Motor neurones – Carry impulses from CNS to effectors.

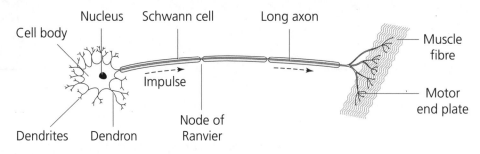

3. **Reflex arcs** are the simplest form of a nervous pathway. They control reflex actions, which are involuntary, rapid responses to a specific stimuli concerned with the protection of the body, e.g. removing hand from painful stimulus. Cranial reflexes are pathways across the brain, involving the reflex actions of receptors near to the brain, e.g. constriction of pupil in eye due to bright light.

Does not involve the conscious brain

4. **Nervous impulses** are chemical and electrical changes along neurone membranes. A difference in charge is set up across the axon membrane – **potential difference** – by movement of Na^+/K^+ ions through axon/dendron membrane proteins at the nodes of Ranvier.

When no impulse is present (**resting potential**) Na^+ ions are actively transported out of the axoplasm into the tissue fluid, K^+ ions are actively transported into the axoplasm out of the tissue fluid. Because the membrane is permeable to K^+ ions, 50% diffuse back in – membrane is **polarised** (-70 mV).

Differences in concentrations of ions cause differences in charge

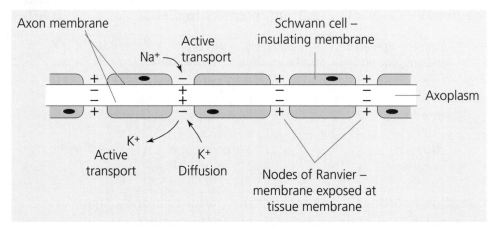

5 A stimulus causes the membrane to become permeable to Na+ ions, causing them to diffuse, in reversing the charge difference. The membrane is **depolarised** causing an action potential (^+30mv). The charge difference between two nodes of Ranvier causes a localised electric current, stimulating the next to become depolarised. The original node of Ranvier returns to a resting potential. **Schwann cells** cause the impulse to flow over a wider area, speeding up transmission.

To generate an impulse a stimulus has to exceed a certain **threshold level**. Impulses are always of the same strength and speed, but the stronger the stimulus the more impulses are generated per second.

'All-or-nothing response'

6 A synapse is a microscopic gap between one neurone and the next, too wide for electrical impulses to cross. Impulses arriving at the end (**synaptic knob**) cause synaptic vesicles to move to the end of the membrane (**pre-synaptic membrane**) due to an influx of Ca^{2+} ions. This releases a transmitter chemical, which diffuses across the gap and binds to receptor sites on the next neurone (**post-synaptic membrane**). A new impulse is stimulated by Na+ ions moving in.

Ca^{2+} ions move through membrane proteins

Transmitter chemical	Location in nervous system	Enzyme responsible for breakdown
Acetylcholine	• voluntary (all) • autonomic (parasympathetic)	cholinesterase
Noradrenaline	• autonomic (sympathetic)	monoamine oxidase

7 Post-synaptic potentials (PSPs) produced can be in two forms:

- **Excitatory (EPSP)** – Neurotransmitter depolarises the post-synaptic neurone, by causing an influx of Na+ ions through membrane channel.
- **Inhibitory (IPSP)** – Neurotransmitter opens K+ or Cl- channels in the post-synaptic membrane. Outflowing of K+ or inflowing of Cl- ions makes the membrane more negative – **hyperpolarisation**. This prevents depolarisation, as the membrane potential is even further from threshold. The cell is less likely to generate an action potential.

8 The cerebral hemispheres in the brain send messages to the cerebellum region which controls the muscles.

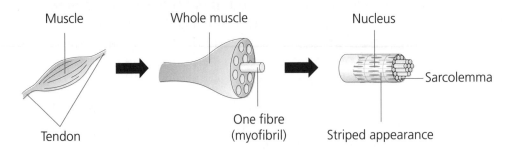

Muscle   Whole muscle   Nucleus

Tendon

One fibre
(myofibril)

Striped appearance

Sarcolemma

The **myofibrils** making up the muscle are composed of proteins – **actin** and **myosin**, and two membrane systems: 1. sarcoplasmic reticulum. 2. T-systems connecting the sarcolemma.

Membrane system covers entire myofibril

Actin is a thin filament consisting of two helical strands of globular proteins. Actin molecules contain a myosin binding site, and two accessory proteins:

- **Tropomyosin** – rod-shaped protein wrapped around actin.
- **Troponin** – a globular protein with three subunits which control muscle contraction. The first links tropomyosin, second binds calcium ions and third initiates interaction between actin and myosin.

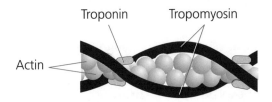

Troponin   Tropomyosin

Actin

Myosin is a thick filament made of long fibrous proteins with globular heads on a flexible hinge. The heads form attachments with actin filaments.

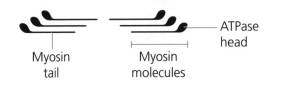

Myosin
tail

Myosin
molecules

ATPase
head

9 **Contraction** is stimulated by nerve impulses arriving at **neuromuscular junctions**:

- Acetylcholine binds to receptors on the sarcolemma causing depolarisation.
- Depolarisation spreads to the sarcoplasmic reticulum, along T-tubules, causing release of Ca^{2+} ions.
- Ca^{2+} ions bind to troponin, moving troponin and tropomyosin, exposing the myosin binding site.
- ATP binding to myosin causes the molecule to change shape and bind to the binding site on myosin.
- ATPase breaks down ATP causing a second shape change in myosin and the energy released pulls the actin towards the M line. The sarcomere is shortened.

Actin and myosin forms cross bridges

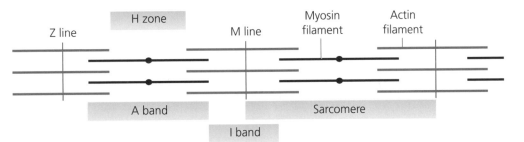

Region	Description	Muscle contraction
M line	Centre of actin	No change
Z line	Separates two sarcomeres	No change
I band (lightest)	Contains only actin filaments (+ Z line)	Decreases (overlap increases)
H zone (darker)	Myosin filaments only	Decreases (overlap increases)
A band (darkest)	Length of myosin filament (contains action + myosin)	Remains the same length

You must know how a sarcomere changes during contraction

Now learn how to use your knowledge

Use your knowledge

40 minutes

Hints

1 The diagram below shows a sensory neurone.

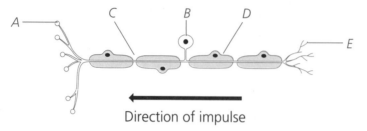

Direction of impulse

(a) Label parts *A* to *E* (5).
(b) Explain the role of the Schwann cells in the neurone.

2 The diagram below shows an electron micrograph of a longitudinal section of myofibril in striated muscle.

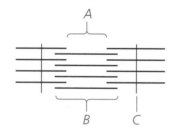

(a) Name the regions *A*, *B* and *C*.
(b) State the name of the two proteins present in the myofibrils.
(c) Explain the role of the following in muscle contraction:
   (i) ATP.
   (ii) Proteins present in the myofibrils.
(d) State how the banding patterns in the myofibril will change during muscle contraction.
(e) Describe how muscle contractions are stimulated at the neuromuscular junction by the arrival of nervous impulses.

Remember filaments slide past each other

✓ *Answers on page 93*

Reproduction

Test your knowledge

15 minutes

1 Reproduction may be sexual or _____. Sexual reproduction involves _____ parents whereas asexual reproduction always involves one parent. The offspring from asexual reproduction are _____ to their parents and each other.

2 Pollination is the transfer of _____ from the _____ to the stigma. Insect pollinated flowers have large _____ petals, which are usually scented, and have a small, sticky _____ to hold pollen.

3 At fertilization, one pollen nucleus fuses with the _____ _____ producing a zygote which develops into the _____. The second pollen nucleus fuses with the diploid cell, producing a _____ nucleus which develops into _____ tissue.

4 Spermatogenesis is the production of spermatozoa and is triggered by the hormone _____.

5 At puberty, females release one egg cell – a secondary _____ – from alternate ovaries per _____.

6 FSH stimulates the development of a _____ _____ in an ovary and stimulates it to release oestrogen. In anticipation of fertilization oestrogen promotes _____ of the endometrium. Ovulation is triggered by _____ hormone on day _____. If fertilization occurs the blastocyst secretes _____ which maintains the corpus luteum and therefore maintains levels of _____ and _____.

7 To enable fertilization the _____ in the head of the sperm bursts releasing _____ which digest the outer layer of the egg.

8 Birth is triggered by oxytocin from the _____ _____ gland and prostaglandins from the _____.

Answers

1 asexual, two; identical 2 pollen, anther, coloured, stigma
3 egg cell, embryo, triploid, endosperm 4 FSH 5 oocyte,
ovulation 6 Graafian Follicle, thickening/vascularisation, luteinising,
14, HCG, progesterone, oestrogen 7 acrosome, enzymes
8 anterior pituitary, placenta

✓ *If you got them all right, skip to page 78*

73

Reproduction

Improve your knowledge

20 minutes

1 Reproduction may be **sexual** or **asexual**.

Asexual: e.g. bacteria	Sexual: e.g. animals and higher plants
One parent only	Usually two parents, as in algae, fungi, coelenterates and animals and higher plants
No gametes produced	Gametes produced: small, mobile, male gametes and large, stationary female gametes e.g. sperm, ova, antherozoids, oospheres
Mitosis only involved	Mitosis and meiosis involved. Meiosis usually to produce haploid gametes
Offspring identical to parents and each other, except when mutations occur	Offspring varied from parents and each other because of mutations, crossing over and independent assortment in meiosis, combination of parental genomes and random fertilization. Variation allows evolution
Usually many offspring, produced very rapidly	Usually fewer offspring, produced more slowly

The key point is that sexual reproduction always produces offspring which are different from the parents. This **variation** allows the process of **natural selection** to operate and hence organisms **evolve**.

2 Pollen is produced in the stamens, which consist of an **anther** and a long **filament**. The transfer of pollen from the anther to the stigma is called **pollination**. There are many differences between wind- and insect-pollinated flowers.

Wind-pollinated flower	Insect-pollinated flower
Small petals (usually green) or petals absent; flowers inconspicuous	Large coloured petals; conspicuous flowers/inflorescences
Not scented	Scented
Nectaries absent	Nectaries present
Large branched and feathery stigma hanging outside flower Pendulous stamens hanging outside flower to release pollen	Small sticky stigma to hold pollen, usually enclosed within flower Stamens enclosed within flower
Large quantities of pollen Pollen grains relatively light and small; dry, often smooth walls	Less pollen produced Pollen grains relatively heavy, large, possessing spines or being sticky for attachment to insect body
Simple flower structure	Complex flower structure

3 After being deposited on the stigma, the pollen grain produces a pollen tube which goes down the style and eventually enters the ovule via a tiny hole called the **micropyle**. The generative nucleus of the pollen grain forms two sperm nuclei. These pass down through the pollen tube to the female egg cell. **One** of the sperm nuclei fuses with the egg cell, producing a **zygote** which develops into the **embryo**. The **second** sperm nuclei fuses with the diploid central cell, producing a **triploid nucleus** which develops into **endosperm tissue** which provides food for the developing embryo.

4 **Spermatogenesis** is the production of spermatozoa. It occurs in the seminiferous tubules of the testes and is triggered by **FSH** from the anterior pituitary.

		Chromosome number
Germinal epithelial cell	(2n)	46
Spermatogonium	(2n)	46
Primary spermatocyte	(2n)	46
Secondary spermatocyte	(n)	23
Spermatids	(n)	23
Spermatozoa	(n)	23

Mitosis

Mitosis

Meiosis I

Meiosis II

Maturity

5 **Oogenesis** is the production of egg cells and occurs in the **ovaries** of the female foetus. Oogenesis follows similar stages to spermatogenesis. At **puberty**, one egg cell – a **secondary oocyte** – is released per month in ovulation. This egg cell has undergone the first meiosis division but has stopped at metaphase of the second meiotic division. This is completed only if the egg cell is fertilized by a sperm.

6 The female **menstrual cycle** is controlled by hormonal secretion.

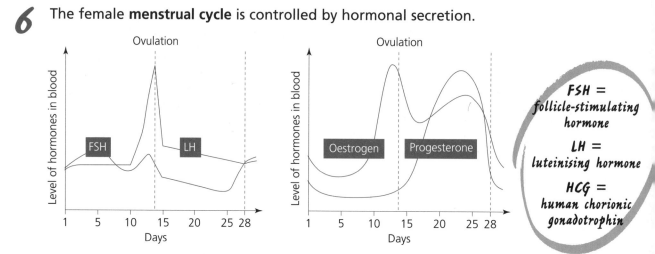

FSH = follicle-stimulating hormone

LH = luteinising hormone

HCG = human chorionic gonadotrophin

- Day 2   Increasing FSH from the anterior pituitary stimulates the development of a Graafian Follicle in an ovary and stimulates the follicle to release oestrogen.
- Day 12 Oestrogen causes the thickening and vascularisation of the endometrium of the uterus, increases LH from the anterior pituitary, and decreases FSH (negative feedback).
- Day 14 Increasing LH leads to ovulation (release of secondary oocyte) from Graafian Follicle.
          Increasing LH and prolactin causes the Graafian Follicle to develop into the corpus luteum.
          The corpus luteum produces progesterone and oestrogen which maintain the endometrium.
          Increasing progesterone leads to a decrease in release of FSH and LH, so no more follicles develop.
- If fertilization occurs the zygote divides mitotically to form a blastocyst, which implants in the endometrium.

The blastocyst secretes HCG, which maintains the corpus luteum and therefore increases progesterone and oestrogen.

- **Week 10** The placenta forms and produces progesterone.
- If **no fertilization** occurs, the corpus luteum stops producing progesterone and oestrogen. FSH production, which had previously been inhibited by high progesterone, now increases again and a new follicle begins to develop. Decreasing progesterone and oestrogen causes the endometrium to disintegrate and it is lost from the body with blood, in the menstrual flow.

7 To enable fertilization the acrosome in the head of the sperm bursts, releasing enzymes which digest the outer layer of the egg. As soon as one sperm has penetrated the egg a fertilization membrane forms, preventing any other sperm entering. The egg now completes a second meiotic division. The sperm and egg nuclei fuse to form a diploid zygote.

8 Birth is triggered by:
- Decreasing progesterone, which had previously inhibited uterine contractions.
- Oxytocin from the anterior pituitary and prostaglandins from the placenta which stimulate contractions.

Lactation involves the production of milk for the baby and is stimulated by prolactin from the pituitary gland.

Now learn how to use your knowledge

Reproduction

Use your knowledge

15 minutes

1 The diagram below shows some of the stages of the development of pollen grains:

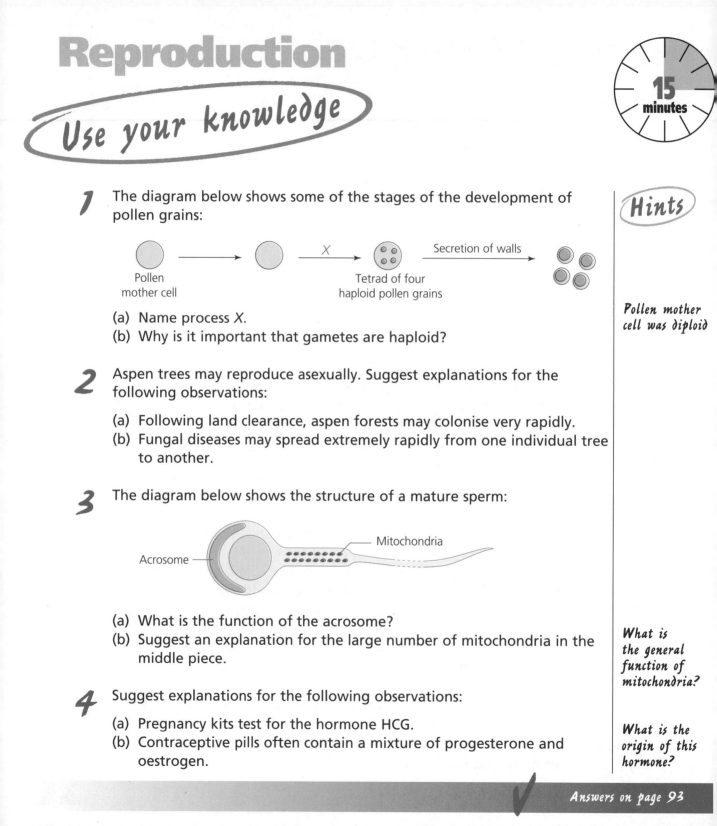

Pollen mother cell

X

Tetrad of four haploid pollen grains

Secretion of walls

(a) Name process X.
(b) Why is it important that gametes are haploid?

Pollen mother cell was diploid

2 Aspen trees may reproduce asexually. Suggest explanations for the following observations:

(a) Following land clearance, aspen forests may colonise very rapidly.
(b) Fungal diseases may spread extremely rapidly from one individual tree to another.

3 The diagram below shows the structure of a mature sperm:

Acrosome

Mitochondria

(a) What is the function of the acrosome?
(b) Suggest an explanation for the large number of mitochondria in the middle piece.

What is the general function of mitochondria?

4 Suggest explanations for the following observations:

(a) Pregnancy kits test for the hormone HCG.
(b) Contraceptive pills often contain a mixture of progesterone and oestrogen.

What is the origin of this hormone?

Hints

✓ *Answers on page 93*

Homeostasis and defence

Test your knowledge

1 Homeostasis involves the process of _____ feedback. If a system moves out of equilibrium, this kind of feedback process will try to return the system to equilibrium.

2 The body temperature of _____ such as insects and reptiles varies with that of the _____.

3 If the blood temperature of a mammal increases this is detected by receptors in the _____ of the brain. The temperature will then be decreased by physiological means such as _____ and sweating.

4 Increased blood glucose concentrations are detected by _____ cells in the _____ _____ _____ in the pancreas which release the hormone _____.

5 Clotting occurs when platelets secrete _____. This converts prothombin into _____. In turn this converts fibrinogen into fibrin.

6 Disease-causing organisms are known as _____. If such organisms are ingested, the enzyme _____ contained in saliva may kill them.

7 Cells which are damaged by scratches or cuts release the chemical _____ which causes vasodilation, leading to swelling. This attracts phagocytes such as _____ which ingest bacteria.

8 In the specific immune response, B-lymphocytes release _____ which 'label' pathogens making it easier for the body to kill them.

Answers

1 negative   2 ectotherms, environment   3 hypothalamus, vasodilation   4 beta, Islets of Langerhans, insulin   5 thromboplastin, thrombin   6 pathogens, lysozyme   7 histamine, neutrophiles   8 antibodies

If you got them all right, skip to page 83

Homeostasis and defence

20 minutes

Improve your knowledge

1. **Homeostasis** is the maintenance of internal body conditions such as temperature, blood pH and blood pressure within narrow limits. This allows cells to function effectively. Often, homeostasis involves **negative feedback**. When a system moves out of equilibrium, e.g. if a mammal's blood temperature increases – feedback processes will begin to restore the equilibrium – in this case to decrease blood temperature.

2. **Temperature control** is vital so that enzymes can work at their optimal rate. The body temperature of **ectotherms** such as reptiles or insects varies with that of the environment. The only way ectotherms can regulate their body temperature is by behavioural means, e.g. marine iguanas in the Galapagos Islands bask on sunny rocks to heat up or lie on damp rocks to cool down.

3. Birds and mammals are **endotherms**. This means that they are able to maintain their body temperature above that of their environment by behavioural and physiological means. The table below shows some of the ways in which a mammal can cool down.

   Endotherms can colonise more areas but have a greater energy demand

Behaviour	Physiological
Flapping ears, e.g. elephants Spreading limbs to increase surface area, e.g. monkeys	Relaxation of hair erector muscles – less air is trapped, more heat is lost by convection
Wallowing in mud, e.g. hippo Water spraying, e.g. elephant	Vasodilation increases blood flow to skin, which increases heat loss by convection and radiation
Panting, e.g. dog Licking fur with excess saliva, e.g. cat	Increase sweating so more heat loss due to latent heat of evaporation

If the **hypothalamus** detects a drop in blood temperature, the opposite processes begin and, in addition, heat is generated by shivering and as a consequence of increased metabolic rate.

Although endothermy allows organisms to exploit cold habitats, one disadvantage is that it **requires a greater food intake** and may necessitate **hibernation** during periods when food is scarce.

4 Careful **control of blood glucose** concentrations is essential.

- Increased blood glucose concentrations are detected by **beta** cells in the Islets of Langerhans in the pancreas, which then release insulin.
- Insulin increases the entry of glucose into cells in the liver, muscles and adipose tissue.
- Inside these cells glucose is converted to glycogen, proteins or fats.
- Decreased blood-glucose concentrations are detected by **alpha** cells, which release glucagon. The glucagon activates enzymes which convert glycogen into glucose, which is then released into the blood. People who suffer from diabetes mellitus cannot control their blood glucose concentrations.

5 Excess blood loss is prevented by **haemostasis**. Damage to a blood vessel immediately results in its constriction, which decreases blood flow and loss. First one platelet and then hundreds more begin to stick to the exposed collagen fibres of the endothelium surface. This blocks blood flow. Clotting then occurs as follows:

- Platelets secrete **thromboplastin**.
- Thromboplastin converts **prothrombin** to **thrombin**.
- Thrombin converts **soluble fibrinogen** into **insoluble fibrin**.
- Fibrin plugs the damaged vessel.

6 The body's first line of defence against **pathogens** (disease-causing organisms) involves the digestive, respiratory and genito-urinary systems and the skin.

Entry of pathogen	Defence mechanism
In food or drink	Lysozymes in saliva digest the pathogen Stomach contains hydrochloric acid
Inhalation	Cilia waft particles to be swallowed Macrophages in alveoli ingest pathogens
Sexual transmission	Acid in mucus Urine flow
Across the skin	Sweat contains lactic acid which kills pathogens Rapid wound healing

7 Scratches, cuts and burns cause a **non-specific immune response**. This induces **inflammation** and **phagocytosis**. Damaged cells release **histamine** which cases vasodilation, increasing blood flow to the area, causing swelling. This attracts phagocytes such as **neutrophils** which ingest bacteria.

8 The entry of pathogens into the body causes a **specific immune response**. This has two components:

- B-lymphocytes produced in the bone marrow release specific antibodies which 'label' the pathogens making it easier for the body to recognise and destroy them. Some antibodies neutralize the toxins produced by the pathogens.
- T-lymphocytes destroy pathogens such as fungi.

Homeostasis and defence

Use your knowledge

1 Suggest explanations for the following observations:

(a) In low temperatures humans develop goose pimples.

(b) People who are overheating appear red-faced.

2 The diagram below shows the mechanisms by which the body controls blood glucose levels.

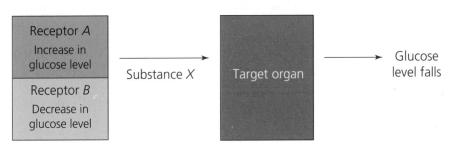

(a) Identify:
   (i) Receptor A.
   (ii) Substance X.
(b) Suggest the name of one target organ which would be influenced by substance X.
(c) Suggest why the process of negative feedback is faster to control body temperature than blood glucose levels.

Hints

What effect does substance X have?

How does the message travel between the receptor and the effector?

Answers on page 94

Test your knowledge

15 minutes

1 A _____ is a group of individuals in an area which can interbreed. They are found in _____ which are communities of organisms interacting with their _____ environment.

2 Populations can increase due to _____ or births. _____ growth occurs when a species colonises a habitat and nothing limits growth.

3 Density _____ mortality factors such as disease mean the proportion of a population dying increases as density increases.

4 In food chains each stage is known as a feeding or _____ level. Transfer of energy between levels is inefficient because energy is _____ at each trophic level.

5 Pyramids of numbers represent the numbers of _____ at each trophic level. Their main disadvantage is that they can be _____, e.g. when based on a single tree.

6 _____ convert dead organic matter into small particles, which aids decomposers to _____ break down large organic molecules.

7 Succession from bare ground is called _____ succession. As the soil develops, smaller plants are replaced with larger plants, until the _____ _____ is reached.

8 _____ is the clearance of woodland, often to create land for crops. The consequences include increased _____ erosion, _____ of rivers and habitat loss.

9 Stratospheric ozone absorbs _____ radiation from the sun. Depletion has occurred during the twentieth century due to release of _____ from aerosols and as coolants.

10 Acidic _____ is the deposition of acidic gases from the atmosphere, produced by combustion of _____ fuels. The consequences include _____ of freshwater.

11 Greenhouse gases allow entry of _____ radiation from the Sun, and _____ loss of re-radiated long-wave radiation.

Answers

1 population, ecosystems, physical **2** immigration, exponential **3** dependent **4** trophic, lost **5** organisms, inverted **6** detritivores, chemically **7** primary, climax community **8** deforestation, soil, silting **9** short-wave, CFCs **10** precipitation, fossil, acidification **11** short-wave, delay

✔ *If you got them all right, skip to page 90*

Ecology
Improve your knowledge

20 minutes

1 A **population** can be defined as a group of individuals in an area which can interbreed. Populations of different species interact together in an area forming a **community**.

An **ecosystem** is any community of organisms interacting with one another, and with their physical environment. Within an ecosystem, organisms live in a particular place – **habitat** – defined by its physical characteristics. The **ecological niche** is the role played by a species in the ecosystem. This includes its habitat, plus its position in the community, e.g. producer, predator.

Ecological niche is the 'job' an organism has in an ecosystem

2 If a population is not changing, the growth rate (change in numbers over time) is zero. Numbers in a population can increase (positive growth rate), due to **births and immigration**, or decrease (negative growth rate), due to **deaths and emigration**.

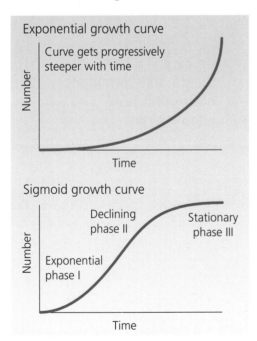

Population is growing at its maximum rate, with births higher than deaths. This occurs when a species colonises a new habitat and no factors, e.g. food, space, light, limit the population growth.

Exponential growth cannot continue. Population number will reach a level where there are too many individuals and **intraspecific competition** (competition between individuals of the same species) occurs for resources, so growth rate declines.

All habitats have a maximum population size that can be supported

85

3 Competition is a major factor determining population size. A sigmoid growth curve shows that intraspecific competition is a **density-dependent** mortality factor. The proportion dying increases as population density increases because the level of competition for resources, e.g. food, space, water, light, increases. Other examples of density-dependent factors are predation and contagious diseases.

Density-dependent means percentage dying increases as density increases

Climate is a **density-independent** mortality factor, with the proportion of population dying independent of the density, e.g. a cold winter will kill individuals of a young bird population, whatever the population density.

4 **Food chains** are a linear sequence of organisms in a feeding relationship. Each stage is known as a feeding or **trophic level**.

Oak tree	→	Earthworms	→	Shrews	→	Owls
Producer	→	Primary consumer	→	Secondary consumer	→	Tertiary consumer
Producer	→	Herbivore	→	Carnivore	→	Carnivore

Arrows indicate energy flow between trophic levels, starting at the producer

Food webs are a system of numerous food chains and are a more realistic representation of feeding relationships, since many organisms feed on more than one species and at more than one trophic level.

There are rarely more than four trophic levels because there is not enough energy to support more. The **transfer of energy** between trophic levels is inefficient, with energy **lost** at each trophic level for several reasons:

- Not all of the organisms at any trophic level are eaten.
- Some organic molecules, e.g. cellulose in plants, are undigestable to some organisms and so not available to the next trophic level.
- Energy is lost at each trophic level as heat due to respiration.

Pyramid of numbers	Pyramid of biomass	Pyramid of energy

Tertiary		90
Secondary		1500
Primary consumers		14 000
Producers		88 000
Number of individuals	Dry mass/g m^{-2}	kJ m^{-2} y^{-1}
Represents number of organisms at trophic level	Total dry mass of organic matter per trophic level (in a unit area or volume)	Organisms are converted to their energy equivalent kJ m^{-2} y^{-1}

Advantages

• Easy to produce • Non invasive (no organisms killed)	• More realistic representation of food web as usually upright	• Always upright, because energy is always lost at each trophic level

Disadvantages

• Can be inverted, e.g. one oak tree at base • Numerous small consumers, e.g. parasites, unbalance the pyramid	• Needs more data to construct because it is based on **dry** mass • Organisms are killed • Can be inverted due to seasonal variations, e.g. variations in phytoplankon populations	• Difficult to construct. Many samples must be taken over a long period and all samples must be killed, dried, weighed and burnt to measure energy content

6 **Detritivores** are animals which feed on dead organic matter **mechanically breaking down** (comminuting) large particles into small ones. **Decomposers** are bacteria and fungi which feed on dead organisms **chemically breaking down** (digesting) large organic molecules into small ones.

Detritivores and decomposers form the **detrital food chain**. This is a vital part of any ecosystem as dead organic material is broken down into its simple inorganic ions making these mineral nutrients available for uptake by plants, i.e. inorganic ions are recycled.

Action of detritivores increases the surface area for decomposers

7 **Succession** is the development of a stable community (the climax community) over time, through a number of stages. This results in a long-term change in the composition of the community due to the modifying effect that organisms have on their environment.

Succession from bare ground is known as **primary succession** and can take hundreds of years. **Secondary succession** occurs where vegetation has already been present, e.g. burnt heathland or cleared woodland, and occurs in a shorter period of time.

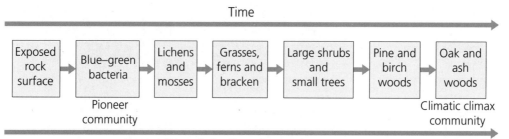

- Algae and lichen form the **pioneer community** and are able to colonise areas with no organic matter, e.g. bare rock from a landslide. They cause the accumulation of organic material and weathered rock, forming a young soil.
- The soil allows mosses, ferns and small herbaceous plants to grow. These species out-compete and replace the pioneer-community species (**interspecific competition**), and more soil accumulates.
- As the soil develops further, interspecific competition results in these small plants being replaced by larger plants, until the **climax community** is reached. This community is in **equilibrium** with the climate. In England the climax community is usually deciduous woodland.

The environment becomes less harsh with succession, so more species can colonise

8 **Deforestation** is the clearance of woodland, often to create land for crop fields, housing or mining and supply wood for fuel. The extent of damage has increased with increasing mechanisation and growing human populations. The effects of deforestation include:

- Increased **soil erosion** and **leaching** of nutrients as rain impact and run-off increases, along with wind erosion.
- Rivers become **silted up** with eroded material and may **flood** due to the increased volume of water reaching them.
- **Habitats** are **lost** for many species, so reducing species diversity.

9 **Stratospheric ozone** (O_3) found between 15 km and 30 km, absorbs **short-wave radiation** from the Sun, protecting living organisms from this radiation which can cause skin cancers and DNA mutation.

Human activity has caused significant thinning of the O_3 layer, particularly at the poles, mainly due to the use of **CFCs** (in aerosols and as coolants) and nitrous oxides (released from artificial fertilizers). When released into the atmosphere these chemicals break down O_3 faster than it is reformed. This has led to **increased cases of skin cancer** and **reduced crop yield** due to damage to vegetation, particularly in the southern hemisphere where greatest thinning has occurred.

10 **Acidic precipitation** is the deposition of acidic gases (pH 4–4.5) from the atmosphere produced by combustion of coal and oil (fossil fuels). The principal gases are sulphur dioxide and nitrogen dioxide.

Gases may be deposited directly onto surfaces (**dry deposition**) or dissolved in water droplets (**wet deposition**), causing varied effects:

- **Decline in forests** due to acid soil conditions and by making heavy metal ions in the soil available for uptake by roots, so poisoning them.
- **Acidification of freshwater**, killing fish and invertebrates.
- **Chemical weathering** of buildings and monuments.

11 The greenhouse effect is the natural warming of the Earth's atmosphere due to greenhouse gases, e.g. CO_2. These gases allow entry of short-wave radiation from the Sun, but delay the loss of re-radiated long-wave radiation, causing an overall warming effect.

Burning of fossil fuels, e.g. at power stations, releases greenhouse gases

Human activities have increased the concentration of greenhouse gases in the atmosphere. This **enhanced greenhouse effect** has been linked with global climate change, including increased temperatures and declining rainfall.

Now learn how to use your knowledge

Ecology

30 minutes

1 Define the following ecological terms:
   (a) Community.
   (b) Pyramid of energy.
   (c) Succession.

2 Natural climax communities are found in many areas of the world.
   (a) What is meant by the term climax community?
   (b) Suggest three factors that can prevent the climax community being reached.

3 Define the following terms:
   (a) A population.
   (b) Intraspecific competition.
   (c) Suggest three factors that may influence the length of time taken for a wild population to reach the carrying capacity.

4 The graph shows the population change over 24 hours in yeast cells following introduction into a large volume of well-aerated nutrient solution.
   (a) Suggest why the population changes between point A and B.
   (b) Suggest two factors that may determine the population's carrying capacity.

5 Energy flow through an ecosystem is linear:
   (a) Explain what is meant by this.
   (b) Explain why most food chains are limited to four trophic levels.

6 The pyramid of biomass represents the dry mass (g/m³) of plant plankton and animal plankton in the North Atlantic.
   Comment on the shape of this pyramid.

Zooplankton
Phytoplankton

Answers on page 94

Answers to

 *Use your knowledge* **tests**

Cells and membranes

1

	Simple diffusion	Facilitated diffusion	Active transport
Is ATP required?	N	N	Y
Rate of movement	Slow	Fast	Fast
Direction of transport in relation to concentration gradient	Along	Along	Against

2 (a) Y = Glycocalyx, X= lipid bilayer.
(b) lipids/proteins can move laterally/exchange places.
(c) (i) Metabolism of lipids.
(ii) Very active/needs lots of ATP.

Enzymes and respiration

1 (a) Active site; steric relationship with substrate; other molecules unable to bind.
(b) Temperature; enzyme concentration; substrate concentration; presence/absence inhibitor; pH.
(c) (i) Site where substrate binds; forming steric relationship with substrate.
(ii) Inhibitor structurally similar to substrate; competes with active site; reversible.

2 (a) Maintain constant pH.
(b) Break open cells to release enzyme/catalase; increase surface area.
(c) Hydrogen peroxide.
(d) Water bath, maintain suitable temperature range; balance, constant mass of potato tissue; thermometer, monitor constant temperature.

3 (a) Glucose.
(b) Cytoplasm.
(c) NAD/FAD are hydrogen acceptors; electrons transferred along carriers; progressively lower energy levels; energy released forms ATP; oxygen is the final acceptor.

Organic molecules

1 (a) Must be part of diet; not synthesised by the body.
(b) Phosphate group present; only two fatty acids.
(c) Carbon, hydrogen, oxygen.

2 (a) Amino acids.
(b) Monosaccharides.
(c) Monosaccharides.
(d) Fatty acids.

3 (a) (i) Condensation; water.
(ii) Peptide.
(iii) Dipeptide.
(b) Sulphur; nitrogen.
(c) Sulphur forms disulphide bridges; may be hydrophobic/hydrophilic in tertiary structure; may be acidic/basic/buffer; may be charged allowing electrostatic interaction; capable of hydrogen bonding.

Protein synthesis and nuclear division

1 (a) (i) A = ribosome; B = tRNA.
      (ii) Peptide bond formation; ATP; ligase enzyme; condensation reaction/loss of water.
   (b) (i) UAG.
      (ii) AGC.

2 (a) Reference to bivalents/homologous pairs; reference to chiasmata formation/crossing over.
   (b) 46; 23; 46.
   (c) Homologous chromosomes pair up/formation of bivalents/synapsis; chiasmata formation/crossing over; genetic material exchange between homologous pair; independent assortment; during metaphase I.
   (d) Independent assortment; variation; production of haploid cells; synapsis occurs/ bivalents; two divisions.

Genetics

1 (a) Gene codes for a characteristic; alleles are different forms of a gene.
   (b) (i) White flowers, oval leaves.
      (ii) White flowers, round leaves.
   (c) (i) wwRr; wwRR.
      (ii) wwrr.

2 (a) S; s; T; t.
   (b) SsTt; SsTt; ST; St; sT; st.
   (c) Ratio approximately $9:3:3:1$; differences are due to random fertilization, also crossing over could lead to recombinants.

Mutation and evolution

1 (a) Changes in chromosome structure/changing the gene sequence; deletion/inversion/ duplication/translocation.
   (b) Alteration of a single base/nucleotide in DNA; substitution/inversion/ insertion/deletion.
   (c) Sequence of three bases/nucleotide on mRNA.
   (d) Genetic/allelic composition of an organism e.g. Aa.

2 (a) Geographical isolation.
   (b) Common ancestor (dogs) evolved– radiated out then isolated in different environments; mutations occur; producing variation; natural selection acts on variation; eventually populations cannot interbreed.

3 (a) (i) Change in chromosome structure; not change in genes.
      (ii) Production of new alleles; caused by point mutation.
   (b) Best-adapted individuals reproduce more; pass on successful alleles to offspring; natural selection.
   (c) (i) Allopatric speciation is due to spatial separation of populations/demes; sympatric speciation is due to reproductive isolation of populations/demes.
      (ii) Ecological isolation; seasonal isolation; behavioural isolation; mechanical isolation; gamete isolation.

Nutrition

1 (a) Grana/thylakoid/quantosomes.
   (b) (i) Photolysis.
      (ii) Passed to PS2.
      (iii) Reduce NADP.

2 (a) (i) Plant A; greater range of wavelengths/blue and red both present.
      (ii) Plant C; green light is reflected/little is absorbed.
   (b) Concentration of CO_2/H_2O availability/humidity/mineral ion availability.

3 (a) (i) Stomach.
      (ii) Duodenum.
      (iii) Wall of the ileum.
   (b) Villi/microvilli; large surface area for absorption/lacteal; transport of fats/ rich capillary network; maintain concentration gradient/crypt of Lieberkuhn; Brunner's glands secrete enzymes/smooth muscle; peristalsis.
   (c) Peristalsis; mixing/movement of food along the gut.

Respiration and gaseous exchange

1 (a) Small organisms have large ratios of surface area to volume for rapid diffusion and short diffusion distance; gases can enter and leave along diffusion gradient. Mammals have small ratio of surface area to volume; long diffusion pathways; skin is impermeable; need internal exchange mechanism that is moist; has large surface area; need ventilation mechanism.

(b) Moist; large surface area; transport mechanism; thin surface.

2 (a) Diffusion from atmosphere; through stomata into leaf air spaces; dissolves in water in moist cell walls; CO_2 diffuses into cells in solution down diffusion gradient.

(b) Lenticels; ruptured areas of bark; roots; root-hair cells with large surface area.

3 (a) Inhalation brings oxygen down trachea and down bronchus; through bronchioles into air sac/space; into alveoli.

(b) Prevents collapse of bronchi when pressure falls during inhalation.

4 (a) One cell thick; short diffusion path; high rate of diffusion.

(b) Allows O_2 to dissolve; enables diffusion.

Removal of waste

1 (a) Urine has greater concentration of urea because water re-absorbed; glucose absent in urine/present in filtrate; PCT re-absorbs; chlorine ion higher concentration in urine.

(b) Plasma contains proteins, absent in filtrate; too large to cross basement membrane; more water in filtrate; absence of proteins.

2 (a) H_2O leaves descending limb via osmosis; descending limb impermeable to salts; Na^+/Cl^- pumped out of ascending limb via active transport; ascending limb impermeable to H_2O.

(b) PCT.

(c) Ultrafiltration exceeds re-absorption; not all glucose can be reabsorbed in PCT.

(d) Water is re-absorbed along with solutes.

(e) ADH was present because concentrated urine produced; water re-absorption in DCT/collecting duct.

Transport in plants and animals

1 (a) Sieve tube elements; companion cells.

(b) (i) Phloem.

(ii) Sucrose; amino acids; auxin.

(iii) Nitrate.

2 (a) I = artery; II = vein; III = capillary.

(b) I has muscular wall, narrow lumen; II has valves; III is one cell thick, has pores, narrow lumen.

(c) Capillaries.

Nerves and muscles

1 (a) A – Synaptic knob/membrane; B – Cell body; C – node of Ranvier; D – Schwann cell; E – pre-synaptic knob/membrane.

(b) Insulates membrane; increases speed of transmission.

2 (a) A = H zone; B = A band; C = m line.

(b) Actin; myosin.

(c) (i) Binds to the myosin head; changes shape of head; allows myosin to bind to actin binding site.

(ii) Myosin head binds to actin binding site; second change in head shape pulls filaments along.

(d) H zone shortens; I band shortens; A band unchanged.

(e) Ca^{2+} enters synaptic knob; vesicles fuse with pre-synaptic membrane; transmitter chemical is released; diffuses to post-synaptic membrane; binds to specific receptor site; causes depolarisation.

Reproduction

1 (a) meiosis.

(b) Maintains ploidy; the amount of genetic material would increase at each generation; ensures homologues.

2 (a) Vegetative reproduction is faster than sexual reproduction; the stems develop from underground root systems.

(b) All root systems may be connected; the fungus may be systemic; all trees are genetically identical.

3 (a) Contains hydrolytic enzymes; to digest the outer layer of oocyte; allows penetration of egg.

(b) Synthesise ATP; to provide energy for swimming.

4 (a) It is secreted by the embryo/blastocyst; if present, woman must be pregnant.
(b) High progesterone and oestrogen inhibits FSH secretion; preventing the development of new follicles.

Homeostasis and defence

1 (a) Hair muscles contract and hairs become erect; traps air; poor conductor/good insulator; reduces heat loss.
(b) Vasodilation of arterioles; increases blood flow near skin surface; increases heat loss by convection and radiation; increased blood flow gives colour.

2 (a) (i) Beta cells in Islets of Langerhans.
(ii) Insulin.
(b) Liver.
(c) Temperature control involves the central nervous system; this is quicker than hormones/blood circulation.

Ecology

1 (a) All organisms/populations in a given habitat.
(b) Diagram showing energy content at each trophic level, within a food chain.
(c) Change in communities over time due to changing environmental conditions caused by community present.

2 (a) End-point of succession; community in equilibrium with environment; no physical change will occur.
(b) Soil salinity; soil pH; human activity.

3 (a) Group of freely interbreeding members of the same species in a habitat at a given time.
(b) Competition within species for food/light/space, etc.
(c) Time taken to reach sexual maturity; gestation period; number per litter.

4 (a) Increased competition; build-up of toxic waste.
(b) Food availability; competition; space.

5 (a) Energy flows in one direction from primary producer upwards.
(b) Energy lost at each trophic level; insufficient energy remains.

6 Inverted; represents standing crop; phytoplankton population fluctuates greatly.